DNA AND RNA: PROPERTIES AND MODIFICATIONS,
FUNCTIONS AND INTERACTIONS,
RECOMBINATION AND APPLICATIONS

TELOMERASE

COMPOSITION, FUNCTIONS AND CLINICAL IMPLICATIONS

DNA AND RNA: PROPERTIES AND MODIFICATIONS, FUNCTIONS AND INTERACTIONS, RECOMBINATION AND APPLICATIONS

Additional books in this series can be found on Nova's website under the Series tab.

Additional E-books in this series can be found on Nova's website under the E-book tab.

DNA AND RNA: PROPERTIES AND MODIFICATIONS,
FUNCTIONS AND INTERACTIONS,
RECOMBINATION AND APPLICATIONS

TELOMERASE

COMPOSITION, FUNCTIONS AND CLINICAL IMPLICATIONS

AIDEN N. GAGNON
EDITOR

Nova Science Publishers, Inc.
New York

Copyright © 2010 by Nova Science Publishers, Inc.

All rights reserved. No part of this book may be reproduced, stored in a retrieval system or transmitted in any form or by any means: electronic, electrostatic, magnetic, tape, mechanical photocopying, recording or otherwise without the written permission of the Publisher.

For permission to use material from this book please contact us:
Telephone 631-231-7269; Fax 631-231-8175
Web Site: http://www.novapublishers.com

NOTICE TO THE READER

The Publisher has taken reasonable care in the preparation of this book, but makes no expressed or implied warranty of any kind and assumes no responsibility for any errors or omissions. No liability is assumed for incidental or consequential damages in connection with or arising out of information contained in this book. The Publisher shall not be liable for any special, consequential, or exemplary damages resulting, in whole or in part, from the readers' use of, or reliance upon, this material.

Independent verification should be sought for any data, advice or recommendations contained in this book. In addition, no responsibility is assumed by the publisher for any injury and/or damage to persons or property arising from any methods, products, instructions, ideas or otherwise contained in this publication.

This publication is designed to provide accurate and authoritative information with regard to the subject matter covered herein. It is sold with the clear understanding that the Publisher is not engaged in rendering legal or any other professional services. If legal or any other expert assistance is required, the services of a competent person should be sought. FROM A DECLARATION OF PARTICIPANTS JOINTLY ADOPTED BY A COMMITTEE OF THE AMERICAN BAR ASSOCIATION AND A COMMITTEE OF PUBLISHERS.

Library of Congress Cataloging-in-Publication Data

Telomerase : composition, functions, and clinical implications / editor, Aiden N. Gagnon.
  p. ; cm.
 Includes bibliographical references and index.
 ISBN 978-1-61668-957-5 (hardcover)
 1. Telomerase. I. Gagnon, Aiden N.
 [DNLM: 1. Telomerase--physiology. 2. Neoplasms--enzymology. 3. Neoplasms--genetics. 4. Telomerase--genetics. 5. Telomerase--metabolism. 6. Telomere--physiology. QU 56 T272 2010]
 QP606.T44T447 2010
 616.99'40642--dc22
                    2010014110

Published by Nova Science Publishers, Inc. ✢ New York

Contents

Preface		vii
Chapter I	Telomerase beyond Telomeres: New Roles for an Old Enzyme *Gabriele Saretzki*	1
Chapter II	Telomerase and Telomeric Proteins: A Life Beyond Telomeres *Ilaria Chiodi, Cristina Belgiovine and Chiara Mondello*	35
Chapter III	Histone Deacetylase Inhibition as an Anticancer Telomerase-Targeting Strategy *Ruman Rahman*	59
Chapter IV	Telomerase Role in Pituitary Adenomas *A. Ortiz-Plata*	81
Chapter V	Regulations of Telomerase Activity and *WRN* Gene Expression *Fumiaki Uchiumi, Yoshikazu Higami and Sei-ichi Tanuma*	95
Chapter VI	Telomerase and Telomere Dynamics in Cancer: Clinical Application for Cancer Diagnosis *Eiso Hiyama and Keiko Hiyama*	105
Index		145

Preface

Telomerase is a ribonucleoprotein enzyme that catalyzes the cellular synthesis of telomeric DNA during cellular division, resulting in maintenance of telomere length and increased proliferative potential.

Several studies suggest that the telomerase may play an important role in the diagnosis and prognosis of cancer because its expression strongly correlates with the potential tumor progression. Ninety percent of human cancers on different organs have shown high telomerasa activity. This book reviews research in the field of telomerase including functions of telomerase independent of its interaction with telomeres on gene expression and chromatin structure; histone deacetylase inhibition as an anticancer telomerase-targeting strategy and others.

Chapter I - Since its discovery 1985 in unicellular ciliates by Greider and Blackburn, telomerase has gained increasing attention culminating in the Nobel Prize award for Medicine 2009. Telomerase is a very ancient, evolutionary conserved eukaryotic enzyme that seems to have diversified its functions during evolution.

Two main components of telomerase --TERT (the catalytic subunit) and TR/TERC (the RNA component) -- are necessary and sufficient for the catalytic synthesis of telomere DNA, but many regulatory factors associate with the enzyme as well.

Soon after telomerase was discovered in humans it attracted major interest for cancer research, due to its tight correlation to cellular immortality and the infinite proliferation capacity of tumour cells. This important role is based on the most prominent and best studied (canonical) function of the enzyme which is the maintenance and elongation of telomeres.

However, in recent years evidence has accumulated for functions of telomerase that are independent of its interaction with telomeres. These new functions have been described inside and outside the nucleus and can be independent of catalytic activity or the association of TERT with TR.

TERT on its own can influence gene expression and chromatin structure and interfere with transcriptional regulation of certain signalling pathways during stem development. In addition, it can modulate various cellular functions such as stress resistance, sensitivity against DNA damaging agents, apoptosis induction and even brain function.

An important, yet unexpected finding was the ability of mammalian telomerase to shuttle in and out of the nucleus and into other subcellular compartments such as mitochondria upon oxidative stress.

The biological role of mitochondrial telomerase is not entirely clear yet and in some cases controversial, but some data suggest that it improves mitochondrial function, decreases cellular oxidative stress and sensitivity against apoptosis induction.

By complexing with other cellular RNAs, hTERT can change its biochemical function from being a DNA polymerase to an RNA polymerase controlling different cellular processes. Not all cellular consequences or molecular mechanisms of these new functions are well understood yet but it seems very obvious that there is more to telomerase than telomere elongation. Since telomere dependent and telomere independent functions might co-exist and potentially overlap within a cell it is still controversial whether influencing stress resistance, DNA damage and repair or apoptosis induction, belongs to the canonical or non-canonical functions of telomerase.

A multitude of different non-telomeric functions have been described so far and it can be predicted that more surprises are in store that will extend to more novel cellular functions of telomerase/TERT. These new functions could potentially have major implications for the biology of tumours, stem cells or age-related diseases. To reveal the underlying mechanisms might provide improved and refined anti-neopleastic tools and strategies in the fight against cancer and ageing.

This chapter will summarise and evaluate what is known to date about non-telomeric functions of telomerase and thereby give an insight into the high functional plasticity of an evolutionary old enzyme.

Chapter II - Telomerase was discovered as the enzyme deputed to telomere maintenance. In human somatic cells, its activity is virtually absent, and cells stop proliferating when telomeres fall below a critical length. In contrast, in the vast majority of tumors, telomerase is active and its activity is associated with telomere preservation and with the acquisition of an unlimited proliferative potential. Despite this well known and consolidated role of telomerase, a large body of evidence indicates that it contributes to cellular physiology independently of telomere length maintenance. Multiple roles of telomerase in cancer development have been highlighted, besides telomere lengthening, among which is gene expression regulation. More recently, the telomerase catalytic subunit has been shown to associate with an RNA molecule different from the telomerase RNA, giving rise to a complex with functions unrelated to telomere metabolism. Besides telomerase, telomeric proteins play a fundamental role in telomere maintenance. In particular, a six protein complex, named shelterin, is stably associated with telomeres and is essential for telomere metabolism. As well as for telomerase, additional roles beyond telomeres have been found for telomeric proteins, in particular for the TRF2 subunit of shelterin. In this review, we will discuss extra-telomeric functions of telomerase and TRF2 in different biological processes.

Chapter III - Aberrant epigenetic regulation of gene expression contributes to tumour initiation and progression. Studies from a plethora of haematologic and solid tumours support the use of histone deacetylase inhibitors (HDACi) as potent anticancer agents. The mechanism(s) of HDACi-induced cancer cell phenotypes are complex and incompletely elucidated. This chapter will discuss erroneous epigenetic regulation of hTERT transcription in cancer cells and propose that alleviation of an improper acetylation-deacetylation balance at the hTERT promoter, is one mode by which HDACi induces anticancer effects. The chapter will conclude with some pertinent questions and future perspectives arising from the recent impetus in HDACi studies, with particular attention to the cancer stem cell therapeutic paradigm.

Chapter IV - Telomeres are DNA-protein structures located at ends of each chromosome, which function is to protect the ends of the DNA double helix. Telomeres are implicated in complex biological processes such as regulation of gene expression, cellular senescence, and tumorigenesis. Telomerase is a ribonucleoprotein enzyme that catalyzes the cellular synthesis of telomeric DNA during cellular division, resulting in maintenance of telomere length and increased proliferative potential. Its activity is directly related with the expression of the catalytic component hTERT (human telomerase reverse transcriptase). Several studies suggest that the telomerase may play an important role in the diagnosis and prognosis of cancer because its expression strongly correlates with the potential tumor progression. Ninety percent of human cancers on different organs have shown high telomerasa activity. In cerebral tumors the telomerase activity has been observed on astrocitomas, multiform glioblastomas, meningiomas, oligodendrogliomas and metastatic cerebral tumors, and these results suggest that telomerase activity may be an important marker of brain tumor malignancy. On the other hand, the telomerase activity has not been detected on benign tumors such as the hemangioblastomas and schwannomas, and its expression and the role it plays on the hypophyseal adenomas has not been cleared yet. The hypophysis adenomas represent 10-15 % of intercraneal tumors. Histologically, they are mostly benign tumors that do not show cellular pleomorfism or mitosis figures. However, in some cases they show rapid growth and can spread to nearby structures, have recurrence, be clinically aggressive and even malignize. In spite of the clinical - pathological analysis and the evaluations of the expression of different molecules, aggressiveness and biological behavior of these "malignant" adenomas has not been found. The expression of the catalytic fraction of the telomerase (hTERT) in this type of tumors has been few analyzed. In some reports the hTERT activity has not been detected. In this short communication is presented the possible role of telomerase in pituitary adenomas. In our experience, hTERT expression correlate with cellular proliferation associated with angiogenesis and hormonal activity. In pituitary adenomas, telomerase detection must be correlating with histopathological, ultrastructural and clinical evaluation, expression of proliferation markers, cell cycle molecules, and over all with hormonal activity. The evaluation of these variables with the modern technical tools in molecular biology, as proteomic analysis, could provide information about their biological behavior.

Chapter V - Telomeres, which are the specific structures at the ends of chromosomes, play an important role in regulating genome stability and cellular senescence. Telomerase is a telomere-elongating enzyme composed of TERT and TERC (TR) that are protein and RNA subunits, respectively. Besides telomerase, factors involved in DNA repair synthesis factors have also been shown to regulate telomere length. Among the DNA synthesis factors, WRN, which belongs to the RecQ helicase family, is implicated in telomere regulation. The mutation of the *WRN* gene causes Werner syndrome, in which, patients show genome instability accompanied with premature aging. These observations imply that senescence is controlled by telomere maintenance factors. However, caloric restriction (CR), or glucose deprivation, extends the life span of various animals. We observed the induction of telomerase activity and *WRN* gene expression in 2-deoxy-D-glucose (2DG)-treated HeLa S3 cells. In this article, we discuss the transcriptional regulation of *WRN* and *TERT* gene expressions comparing their promoter regions.

Chapter VI - Telomerase, a critical enzyme responsible for cellular immortality, is usually repressed in somatic cells except for lymphocytes and self-renewal stem/progenitor cells, but is activated in approximately 80% of human cancer tissues. The human telomerase reverse transcriptase (TERT) is the catalytic component of human telomerase. In cancers in which telomerase activation occurs at the early stages of the disease, telomerase activity and TERT expression are useful markers for the detection of cancer cells. In other cancers in which telomerase becomes upregulated upon tumor progression, they are useful as prognostic indicators. On the other hand, in the ectopic expression of telomerase which may precede cancer transformation, detection of telomerase in noncancerous lesion will become a useful marker detecting high-risk patients for cancer development. Moreover, progressive telomere shortening predominantly usually occurs during early carcinogenesis before telomerase activation. And then, the activated telomerase maintains telomere length stability in almost all cancer cells. Recently, methods for the in situ detection of the TERT mRNA and protein have been developed. These methods should facilitate the unequivocal detection of telomerase activated cells, even in tissues containing a background of normal telomerase-positive cells. Thus, telomere and telomerase is useful biomarker for detecting high-risk patients, early detection of cancer, or cancer progression. This review summarizes the data of various kinds of cancers and discusses clinical application of telomere and telomerase in human cancer.

Chapter I

Telomerase beyond Telomeres: New Roles for an Old Enzyme

Gabriele Saretzki[*]
Institute for Ageing and Health, Newcastle University,
Newcastle upon Tyne NE1, United Kingdom

Abstract

Since its discovery 1985 in unicellular ciliates by Greider and Blackburn, telomerase has gained increasing attention culminating in the Nobel Prize award for Medicine 2009. Telomerase is a very ancient, evolutionary conserved eukaryotic enzyme that seems to have diversified its functions during evolution.

Two main components of telomerase --TERT (the catalytic subunit) and TR/TERC (the RNA component) -- are necessary and sufficient for the catalytic synthesis of telomere DNA, but many regulatory factors associate with the enzyme as well.

Soon after telomerase was discovered in humans it attracted major interest for cancer research, due to its tight correlation to cellular immortality and the infinite proliferation capacity of tumour cells. This important role is based on the most prominent and best studied

---

[*] Corresponding author: Dr. Gabriele Saretzki, PhD, Crucible Laboratory, Institute for Ageing and Health, International Centre for Life, Bioscience Centre, Central Parkway Newcastle upon Tyne NE1 3BZ, United Kingdom, Tel: +44 191 241 8813, Fax: +44 191 241 8810, e-mail: gabriele.saretzki@ncl.ac.uk

(canonical) function of the enzyme which is the maintenance and elongation of telomeres.

However, in recent years evidence has accumulated for functions of telomerase that are independent of its interaction with telomeres. These new functions have been described inside and outside the nucleus and can be independent of catalytic activity or the association of TERT with TR.

TERT on its own can influence gene expression and chromatin structure and interfere with transcriptional regulation of certain signalling pathways during stem development. In addition, it can modulate various cellular functions such as stress resistance, sensitivity against DNA damaging agents, apoptosis induction and even brain function.

An important, yet unexpected finding was the ability of mammalian telomerase to shuttle in and out of the nucleus and into other subcellular compartments such as mitochondria upon oxidative stress.

The biological role of mitochondrial telomerase is not entirely clear yet and in some cases controversial, but some data suggest that it improves mitochondrial function, decreases cellular oxidative stress and sensitivity against apoptosis induction.

By complexing with other cellular RNAs, hTERT can change its biochemical function from being a DNA polymerase to an RNA polymerase controlling different cellular processes. Not all cellular consequences or molecular mechanisms of these new functions are well understood yet but it seems very obvious that there is more to telomerase than telomere elongation. Since telomere dependent and telomere independent functions might co-exist and potentially overlap within a cell it is still controversial whether influencing stress resistance, DNA damage and repair or apoptosis induction, belongs to the canonical or non-canonical functions of telomerase.

A multitude of different non-telomeric functions have been described so far and it can be predicted that more surprises are in store that will extend to more novel cellular functions of telomerase/TERT. These new functions could potentially have major implications for the biology of tumours, stem cells or age-related diseases. To reveal the underlying mechanisms might provide improved and refined anti-neoplastic tools and strategies in the fight against cancer and ageing.

This chapter will summarise and evaluate what is known to date about non-telomeric functions of telomerase and thereby give an insight into the high functional plasticity of an evolutionary old enzyme.

Introduction

Since telomerase in protozoans was discovered by Liz Blackburn and Carol Greider in 1985 [1], it has attracted outstanding interest among scientists, clinicians and the public alike. It was initially praised as the "immortality enzyme" [2] that will solve the eternal dream of humanity to fight death and become the "fountain of youth". But even without fulfilling these imaginations, the discovery of telomerase contributed greatly to our understanding of cell proliferation, senescence, tumourigenesis, genomic stability and cellular immortality. The first described and best analysed and understood function of the enzyme is its telomere elongating and maintaining function that brought the Nobel Prize in medicine to its discoverers [1]. This telomere-interacting function has now become known as the canonical function of telomerase. Telomerase needs two essential components: the catalytic subunit TERT and the RNA component TR/TERC that contains the template sequence for the synthesis of telomeric sequences. While the catalytic subunit is rather conserved among species [3] the RNA component may differ in size and sequence but has a highly conserved secondary structure [4].

While in unicellular organisms the enzyme is constitutively active, in multicellular organisms it is differentially expressed. While it is active in many adult tissues in rodents in humans its expression is downregulated early during development [5]. However, the enzyme is prominently re-activated during tumourigenesis and therefore often associated with cellular immortality and cancer [6].

In addition, it is emerging recently that telomerase has several functions that seem to be telomere-independent. These functions, their biological significance and molecular mechanisms are much less well understood than the telomere-dependent function.

Stewart et al [7] demonstrated that the catalytic subunit TERT promotes tumourigenesis independent of its telomeric function, although the mechanism for this effect is not clear yet.

More recent studies demonstrated an influence of telomerase on chromatin structure, gene expression [8-11] sensitivity to DNA damage, and susceptibility to apoptosis [12-14]. In addition, several groups have demonstrated that telomerase shuttles out of the nucleus and into mitochondria [15-20]. Although the function of mitochondrial telomerase seems controversial, most recent papers on the subject point to a protective function of telomerase within mitochondria and for the general redox balance of cells and improvement of mitochondrial function [19-22].

Apart from a different subcellular localisation, hTERT, the catalytic subunit of telomerase, seems to function on its own without catalytic activity [11, 21, 23] or a different biochemical function altogether by complexing with different RNAs [24, 25].

This chapter aims to summarise our current knowledge on functions and properties of telomerase/TERT that seem independent of its function on telomeres which were termed non-canonical functions of telomerase.

Non-Telomeric Functions of Telomerase

1. Chromosome Healing

Barbara McClintock, who won the Nobel price in 1983 for her discovery of mobile genetic elements, had already found in 1939 [26] that chromosomal ends generated by breakage of dicentric chromosomes during anaphase were able to acquire new ends and telomere function in plant embryo tissues while other tissues such as endosperm did not regain this function. McClintock termed this function without knowing about the existence of telomerase "chromosome healing". The differences in "healing activity" between plant endosperm and embryo tissue can be understood today due to the different levels of telomerase in these tissues [27]. Today "chromosome healing" is defined as a process where telomerase recognises broken chromosome ends independently of its sequence composition and elongates them *de novo* with telomere sequences. The addition of new telomeres to the ends of broken chromosomes had been also described in unicellular organisms such as protozoans [28, 29] and yeast [30].

In humans, chromosome healing seems to be a rather rare event. But it has been described in certain genetic diseases such as an alpha thalassaemia that are associated with terminal deletions at chromosomes 16p and 22q. In both cases telomerase adds telomeric sequences to the break sites thereby stabilising the broken chromosome ends [31-33]. The authors suggested that a small complementary DNA region to the telomerase RNA template at the end of a broken chromosome could be sufficient to prime the healing process *in vivo* and this has also been confirmed in vitro using extracts of human cells [34].

Fitzgerald and colleagues [35] analysed this process in detail in different plant species and classified different categories of healing events according to the last base that has been extended by the enzyme and by the part of the hTR template that has been used for this activity. This biochemical activity is essentially very similar to the biochemical activity of telomere elongation where the substrate is the telomeric sequence. The authors describe three distinct classes of plant telomerases with respect to their healing capacity and primer usage. While the Class I soybean telomerase only elongated DNA primers ending in telomeric nucleotides Class II (from Arabidopsis and maize) and class III enzymes from *Sorghum* efficiently extended nontelomeric sequences by positioning the 3' DNA end at a preferred site on the RNA template. These different classes demonstrate a large variability in the interaction of plant telomerases with their substrate DNA on broken chromosome sites. However, the generation of telomeric sequences is much more precise for class I than for class II and III enzymes. The fact that the three classes of plant telomerases do not seem to have any phylogenetic relationships, suggests a high flexibility in the evolution of the active telomerase site [35].

It does not seem that a similar variety of "healing" functions has been demonstrated for human telomerase and might therefore be a particular property of plant telomerases. In general, the addition of unspecific telomere sequences to DNA double strand breaks could have deleterious effects for the efficacy and accuracy of DNA repair processes. Schulz and Zakian have already described in yeast that cells without Pif1 have a 1000 fold increase in random chromosome healing frequency [36]. Makovets and Blackburn [37] discovered the mechanism for the interaction between Pif1 and telomerase in more detail. They describe how the DNA damage signalling in yeast influences telomerase action at DNA breaks by down-regulating it via Pif1 phosphorylation. Their results suggest that targeting TERT away from double strand breaks should prevent aberrant healing of broken DNA ends by telomerase. Thus Pif1 helps to regulate and coordinate competing DNA end-processing activities and therefore promotes DNA repair accuracy and genome integrity. The higher importance of an effective DNA damage repair in mammals and the fact that unicellular organisms and plants deal differently with genomic restructuring could explain the differential significance of telomerase in processes such as chromosome healing in plants and mammals.

2. Gene Expression, Signalling and Chromatin Status

Telomerase expression is often associated with enhanced cell survival under conditions of cellular stress. Although it is difficult to exclude a superior and more efficient telomere capping in cells with high telomerase levels, another possible mechanism is the regulation of gene expression. These genes include those with a role in proliferation, cell differentiation, induction of apoptosis or expression of antioxidant genes [12, 38-40].

Genes that have been found to be correlated to telomerase expression are growth promoting genes such as EGFR and FGF, genes involved in the regulation of cell cycle such as cyclins D1, G2, metabolism, cell signalling or antioxidant defence [8, 9, 39, 41, 42]. However, there does not seem to be a consistent "hTERT dependent signature" since expression changes seem to vary between different cell types as do phenotypes such as proliferation. While some cell types such as epithelial cells [8] show an increased proliferation after hTERT overexpression, fibroblasts do not seem to show the same effect [20].

Dairkee and co-workers analysed various primary breast tumour derived cell lines and compared those that grew continuously with those that ceased proliferation after few passages. In addition, they transduced the latter ones with hTERT *in vitro* and compared them to the *in vivo* immortalised that all expressed high telomerase levels. Using microarray analysis they identified an hTERT-dependent "immortalisation signature" that included changes to genes coding for mitochondrial components as well as oxidoreductases. The presence of the signature correlated to hormone (ER) status and survival of the patients the cells were initially derived from. They proved the functional relevance of hTERT and the corresponding changes in gene expression by knocking down hTERT around 60% using siRNA and found an increase of oxidative stress and apoptosis as well as a decrease of mitochondrial membrane potential. These data correlate well with functional data using stable over expression of hTERT in human fibroblasts described by Ahmed et al., [20].

Knocking down telomerase in various cancer cell lines resulted in changes of expression of genes involved in angiogenesis, metastasis and cell differentiation [9, 10]. Bagheri and colleagues targeted mouse melanoma cells using a ribozyme against TERC. In clones with a high suppression of TERC expression they detected a changed morphology, reduced growth rates and increased melanin synthesis suggesting cell differentiation [10]. At the same time the cell displayed a decreased invasiveness and metastatic potential *in vivo*. In order to identify downstream pathways controlled by telomerase they analysed gene expression changes in TERC suppressed clones and found in addition to genes involved in proliferation and transcription, metabolic changes in glycolytic pathways.

However, when reduction of telomerase levels is connected to secondary processes such as cell differentiation, senescence or apoptosis, it is difficult to discriminate between a direct effect of telomerase onto gene expression or an indirect influence due to the changed cellular state, which might or might not be dependent on a different telomere capping state. Therefore, it could be that as in the study of Bagheri et al., telomerase suppression induced cell differentiation in human melanoma cells, and that this process secondarily correlates with changes in glycolysis and pigment production.

Other groups have reported an interaction of hTERT with the chromatin status in normal or hTERT overexpressing cells [43, 44]. This interference with nuclear chromatin can have implications at multiple levels and cellular functions such as telomere status, gene expression and DNA damage responses. Sharma and colleagues correlated a faster DNA repair in hTERT overexpressing fibroblasts with a higher amount of NTPs, in particular ATP [44]. Masutomi et al had found small amounts of hTERT and telomerase activity in the S-phase of normal fibroblasts [45] that had been thought to be devoid of the enzyme previously. Chromosome healing, as already discussed previously, is one form of DNA repair that involves telomerase. However, this process is rather rare in mammalian cells. The authors suggest hTERT as a regulator of the DNA damage response pathway through its actions on chromatin structure that is different from its known role on telomeres [43]. Since the low level of telomerase present in normal cells is not sufficient to maintain telomere length, the authors speculate that its role is rather in regulating chromatin status. The authors support their suggestion with the fact that TRF2, an important telomere binding protein localises transiently to DNA damage sites [46] and that in general many DNA repair factors associate with telomeres. Suppression of hTERT expression altered the cellular chromatin structure and impaired DNA damage responses including histone H2A.X phosphorylation. These findings suggest that hTERT expression maintains the overall state of chromatin into a configuration that favours the activation of the DNA damage response while a lack of hTERT diminishes DNA repair. This observation could also explain why, in addition to stably maintaining chromosome ends, cells with higher levels of telomerase, such as embryonic stem cells and cancer cells, have better DNA repair and are more resistant against DNA damaging agents than those cells which have no or only negligible amounts of telomerase.

Steve Artandi's group used an inducible TERT expression system in mouse epithelial skin and progenitor cells and analysed the genome-wide transcriptional response to short term changes in TERT levels [41]. The TERT gene they were using was catalytically inactive and therefore the effects they found are not dependent on telomerase activity or telomere stabilisation [41]. As will be discussed in more detail below, the authors discovered specific changes in Wnt and Myc signalling pathways due to overexpression of TERT. The same group showed a year later that TERT acts as a transcriptional cofactor for the β-catenin transcriptional complex by interacting with BRG1, a chromatin remodelling protein [11]. By interacting with a chromatin remodelling protein, TERT changed chromatin configuration. This is the first study that reveals a molecular mechanism for transcriptional modulation of gene expression by TERT in mammalian cells. These data reveal a surprising new role for TERT as a transcriptional modulator of the Wnt/b-catenin signalling pathway. These unexpected findings could have wide implications for further studies on development, regulation of stem cell properties as well as tumourigenesis.

Since these new TERT functions are dependent on nuclear localisation of telomerase, it is tempting to speculate that nuclear exclusion and mitochondrial localisation has something to do with withdrawing telomerase under certain cellular conditions, such as stress, from its nuclear interference, be it telomeric or chromatin regulated. Similar suggestions have been repeatedly made for the nucleolar localisation of telomerase [47]. Alternatively, it is possible that TERT has similar functions in regulating transcription in mitochondria since it has been shown by Judith Haendeler's group that TERT directly binds to several mitochondrial genes [19].

3. Apoptosis, Stress Resistance, Cellular Survival

An altered sensitivity against apoptotic stimuli was among the first property that has been connected to telomerase in a telomere independent manner.

Mark Mattson's group pioneered these observations by demonstrating a correlation of apoptosis induction to either increased or decreased expression of TERT in neurons [48-50]. His group described a protection of neurons by TERT against amyloid beta and showed a decreased sensitivity and better cell survival of PC12 cells overexpressing TERT against etoposide and camptothecin [48, 49]. Another group confirmed these data using neurons from mTERT over expressing mice and showed an increased resistance of these mice against ischemic brain injury and NMDA receptor mediated excititoxicity [21, 51]. Contradictory data exist about the effects of hTERT overexpression on the sensitivity of those cells against apoptosis. While Santos et al., [15, 16] claimd that hTERT overexpressing cells promote apoptosis by hTERT shuttling to mitochondria other groups have found that hTERT overexpression prevents apoptosis induction after oxidative stress or drug treatment [12, 19, 20, 52].

Likewise, cancer cells with compromised telomerase expression show an enhanced sensitivity against various chemotherapeutic drugs and apoptosis [53-57]. Most of these studies report no change in telomere length and a very fast mechanism of apoptosis induction [53, 56, 57].

However a contribution of telomere length or telomere capping for the sensitivity of cells against drugs and oxidative stress cannot totally be ruled out [58]. Since telomerase can exert its action via telomere-dependent and telomere-independent mechanisms, both processes most likely overlap. However, experiments from Lee et al., [51] using an HA-tagged wt shows exactly the same protective effect against induction of apoptosis by staurosporin and wt TERT with no cytochrome c-release from mitochondria or caspase 3 activation that are both hallmarks of apoptosis They found that even transient overexpression of HA-TERT exerts the same anti-apoptotic effect as does a direct inhibitor of apoptosis XIAP. HA-TERT is catalytically active *in vitro* but cannot elongate telomeres [59]. It is tempting to speculate that a direct interaction of the TERT protein with mitochondria, perhaps by preserving a higher membrane potential as shown by Kang and co-workers [21] and Ahmed et al., [20] could be an underlying mechanism for the anti-apoptotic function of telomerase. This connection will be discussed in more detail further-on.

Several groups looked closer into the mechanism of how telomerase protects cells from apoptosis. The inherent interaction of telomerase with the mitochondrial apoptosis pathway has also been confirmed by Massard et al., [13]. They demonstrate that short term depletion of hTERT did not interfere with viability or proliferation of various cancer cell lines. However, it sensitises those cells to apoptotic cell death by various compounds such as cisplatinum, etoposide, mitomycin C and ROS but not to cell death by Fas ligand induction. These authors found an interaction of hTERT with pro-apoptotic molecules such as Bax. Upon depletion of hTERT, Bax got activated by conformational change due to genotoxic drugs while overexpression of Bcl-2 or Bax inhibition prevented apoptosis after drug treatment [13].

Del Bufalo and colleagues [14] analysed the interaction of telomerase with the anti-apoptotic molecule Bcl-2. They found that treatment of cells with the Bcl-2 inhibitor HA14-1 induced shuttling of hTERT from the nucleus to the cytoplasm. While overexpression of hTERT blocked Bcl-2-dependent apoptosis, abrogated mitochondrial dysfunction and nuclear translocation hTERT, downregulation of hTERT using siRNA significantly increased apoptosis if Bcl-2 levels are decreased. They also demonstrated that catalytic activity of TERT was not required for this process [14]. This suggests that it is the TERT protein rather than functional telomerase seems to be responsible for the protection of mitochondria and decrease of apoptosis and therefore excludes the involvement of telomeres in this process. Luiten et al [60] confirmed a decrease of Bcl-2 expression and lower caspase-3 expression in T-cells over expressing hTERT.

Since the nuclear and mitochondrial functions of telomerase can overlap within a cell it is preferable to use experimental systems where telomerase/TERT is confined to specific cellular compartments. However, even under these conditions contradictory results have been reported: Heandeler et al. reported that confining hTERT to the nucleus by using either an exclusion deficient mutant or a nuclearly overexpressed TERT decreased apoptosis induction in HEK293 cells while the same group demonstrated later that mitochondrially targeted TERT in HUVECs was more protective against oxidative stress-induced apoptosis than wtTERT that is expressed in the whole cells [17, 19]. Therefore it is possible that both locations contribute to apoptosis resistance.

So far there exist contradicting data on whether cellular protective effects of telomerase require telomerase activity or not. While some experiments show that telomerase activity is not necessary [61, 62] others claim that activity is necessary for the protective effect since catalytically inactive mutants seem to fail the effect [12, 19]. It is possible that for some cellular function of telomerase its catalytic activity is important while others might not require it. For more details on these specific subclasses of non-canonical function of telomerase the reader is referred to a recent review from Parkinson et al. [63].

Most studies use TERC knockout models in order to analyse telomere dependent defects in later generations. Only very few studies exist on early defects in telomerase knock-out mice [64] that are not dependent on telomere-related effects but rather analyse the effects of the absence of telomerase activity. Due to the possible role of the TERT protein which is still present in TERC knockout mice, valuable information can be gained from the comparison of effects of wild type mice, TERC knockouts and TERT knockout defects in early generation [51]. This unique genetic model is an excellent system in order to discriminate between effects of telomerase including its catalytic activity and the activity independent roles of the two main components of telomerase; the RNA component TERC and the protein moiety TERT.

Lee et al., [51] compared the sensitivity of wt, TERC and TERT knockout mice (first generation) regarding their sensitivity staurosporine and NMDA. While TERT -/- MEFs were highly sensitive against apoptosis induced by staurosporine, TERC-/- mice did not show any different sensitivity compared to wt. Moreover, over expression of hTERT in wt MEFS (which quenches the endogenous telomerase activity in a dominant-negative fashion) [65] further increased the resistance of wild type mice against staurosporine. These results suggest that it is indeed the catalytic subunit TERT that is responsible for the protective effect, even without any catalytic activity.

Similarly, cell death by NMDA was prevented in primary neurons by TERT, but not TERC, both *in vitro* and *in vivo* [51]. As previously, TERT knockout mice showed a decreased survival after NMDA treatment while TERC knockouts did show the same sensitivity as wild type littermates. In addition, the authors applied another, more physiological treatment using sciatic nerve axotomy in motor neurons and found again that a transgenic expression of hTERT in wt MEF protected the neurons from apoptosis [51]. These results show for the first time that the loss of TERT seems to have more dramatic consequences for cell survival after stress than TERC loss and that for many protective functions telomerase activity might not be important, but could derive from TERT protein alone.

4. TERT and the Brain

Various studies suggest an important role of telomerase in protecting neuronal cells from cell death both *in vitro* and *in vivo* [21, 49-51, 66]. While TERT over expression protected pheochromocytoma cells from apoptosis caused by amyloid treatment or trophic factor withdrawal [49, 50] suppression of TERT increased the sensitivity of primary neuronal cells to amyloid as well as excitotoxins such as NMDA (N-methyl-D-aspartic acid) [49, 50]. Kang et al. demonstrated that overexpression of TERT in transgenic animals conferred resistance of mouse brains to ischemic injury and NMDA receptor-mediated neurotoxicity [21].

In the brain telomerase activity corresponds mainly to the subventricular region that contains neural stem cells [67]. However, it emerges that apparently TERT can act on its own in the brain without any catalytic activity. Kang and colleagues did not detect any TERC component or telomerase activity after overexpression of TERT in mouse neurons [21]. However, they still found a protective effect of TERT on NMDA sensitivity, cell death and mitochondrial membrane potential [21].

The same group analysed the significance of telomerase for maintaining normal brain functions *in vivo* [68]. TERT and TERC expression was observed using in situ hybridisation techniques in the cerebellum. However, no coordinated expression of both telomerase components was found in other brain regions such as the hippocampus and olfactory bulbs which lacked TERC expression. These data suggest that TERT but not TERC is constitutively expressed in the hippocampus and the olfactory bulbs. While TERT-deficient mice showed the same cerebellum dependent behaviour as wild type mice they displayed a significantly altered anxiety-like behaviour in a-maze test during ageing [68]. The authors also found a significantly changed olfactory perception in TERT-/- that did not change with age, emphasising the significance of TERT/telomerase for age- independent olfactory function in mice. These changed biological brain functions corresponded to the deviated gene expression patterns observed in the hippocampus and the olfactory bulb brain regions in wild type mice. While most parts of the brain are thought to be post-mitotic, these two regions show a considerable cell turnover that correlates with a supply of stem and progenitor cells from the stem cell generating brain regions such as the subventricular zone that persist throughout adult life. These results suggest that TERT has an important telomere-and telomerase- independent novel role in maintaining the normal behaviour and brain functions.

As already mentioned above, TERT deficiency renders neuronal cells more susceptible to stresses including NMDA-induced neurotoxicity, but TERC deficiency has little effect on this phenomenon [51]. In addition, there is accumulating evidence that apoptotic protection by TERT prevents neuronal cell death (49-51] and involves the suppression of the mitochondrial death pathway [12, 13, 19]. Kim et al., [69] observed that Bax deficient mice have normal olfactory function but a significantly reduced apoptosis rate in the olfactory bulbs. These data suggest that Bax might have an important role in the elimination of olfactory neurons. Thus, Lee et al [68] hypothesise that the mechanism for the phenotypic manifestation of an altered anxiety-like behaviour could be the function of TERT as a major regulator of apoptosis in the brain that supports normal neurogenesis and brain function independent of telomerase activity or telomere effects.

5. Sub Cellular Shuttling of Telomerase

hTERT contains various motifs that allow it to be transported to different subcellular locations. These include nuclear and nucleolar import signals, a nuclear export signal [70] and a mitochondrial localisation signal [15]. The mitochondrial localisation signal seems to be a recent acquisition of higher eukaryotes since only mammals have it while plants have a very fragmented motif and yeast are lacking it [15] although intracellular trafficking of telomerase and its components has been described in yeast [71, 72]. Localisation of telomerase to the nucleolus has been described in humans [73] and yeast [71]. Cytoplasmic localisation has been confirmed for TERT [15, 18-20] as well as for the RNA component [71, 74]. Therefore, shuttling of telomerase between different cellular compartments seems to be a natural and highly regulated process. Nuclear export of telomerase has been described by Seimyia et al as a CRM1/ exportin and 14-3-3-dependent process [70]. 14-3-3 proteins act as molecular chaperons that modify hTERT at a post-translational level and control nucleo-cytoplasmic traffic of hTERT by counteracting the nuclear export of TERT. CRM1 probably binds to the nuclear export signal since deletion of that sequence or inhibition of the nuclear export machinery with LeptomycinB prevents the nuclear export [70]. In T- lymphocytes, telomerase activity is not regulated by the level of hTERT protein. hTERT resides in the cytoplasm until the cells are stimulated by antigen contact leading to the phosphorylation, nuclear import and thereby activation of telomerase [75]. It is not known whether this cytoplasmic localisation is in specific organelles such as mitochondria. However, observations from Haendeler et al. have shown that telomerase can be detected in the cytoplasm in addition to mitochondria [19].

Only in the nucleoplasm can telomerase fulfil any telomere-dependent functions while for any non-telomeric functions telomerase can be within or outside the nucleus. Haendeler and co-workers have shown that upon oxidative stress nuclear hTERT gets phosphorylated by nuclearly localised src kinase leading to an association of TERT with a Ran GTPase [17]. This results in a nuclear export of TERT through nuclear pores. The oxidative stress can be exogenously applied or endogenous, e.g. occuring during the senescence of human endothelial cells [18]. Treatment of cells with antioxidants can prevent the nuclear export, but that could be a secondary process that rather prevents the cellular senescence that depends on ROS generation [76]. The authors found that inhibiting the nuclear export of TERT increases the anti-apoptotic activity of TERT after stimulation with staurosporine and TNFalpha [17]. This would suggest that nuclear telomerase is responsible for its anti-apoptotic properties. This is at odds with results from the same group that demonstrated that mitochondrially localised TERT decreases apoptosis after H_2O_2 treatment in HUVECs [19]. Possibly both localisations contribute to the anti-apoptotic properties-versus a telomere dependent one depending on nuclear TERT and a mitochondrial one that decreases intracellular oxidative stress. A summary of the shuttling process after oxidative stress is given in figure 1.

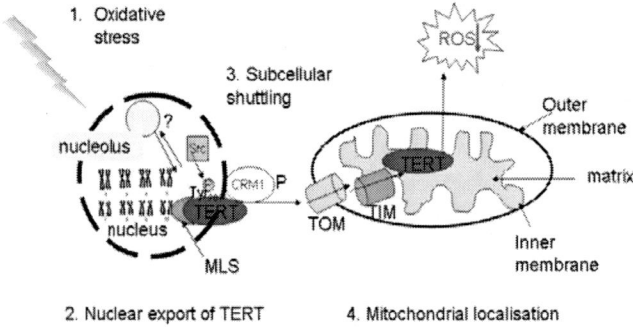

Figure 1. Shuttling of telomerase within a cell upon oxidative stress.

6. Mitochondrial Telomerase

Although it was already known that telomerase can shuttle between different cellular compartments, it came as a surprise to the scientific community when Janine Santos published her paper about an N-terminal mitochondrial localisation site and mitochondrial localisation of telomerase following oxidative stress [15]. That corresponded well with data from Haendeler and colleagues [17] who had described the nuclear exclusion of TERT under oxidative stress. Mitochondria are thought to be endosymbiontic, therefore bacterial in origin and having a circular genome in mammals. It recently has been shown that many proteins that have been found in the nucleus previously can shuttle to mitochondria and perform sometimes different functions from the ones they have in other cellular compartments [77]. Examples are p53, HMGA, prohibitins, Lon proteases, various kinases and many more [77-79]. Some of them bind to components of the mitochondrial respiratory chain and modulate activity of various respiratory complexes [19, 77].

The function of telomerase within the mitochondria is controversial. While Santos et al., [15, 16] found an increased sensitivity for apoptosis inducing agents of cells overexpressing telomerase, most groups describe a decreased apoptosis induction in cells overexpressing telomerase [17, 19, 20, 52] while inhibiting telomerase in cancer cells seems to increase sensitivity against various apoptosis inducing drugs [19, 53, 55-57]. Overexpression of telomerase in various cell types decreases oxidative stress [18-20], increases the mitochondrial membrane potential [20, 21], and increases calcium storage capacity [21, Saretzki, unpublished results, Figure 2]. Haendeler's group demonstrated recently that the import of TERT into the mitochondria uses TOM and TIM proteins [19]. The same group also showed specific binding of TERT to some mtDNA encoded complex I genes, increased oxygen consumption and enhanced complex I activity in cells overexpressing TERT, while mouse lung fibroblasts and heart tissue from TERT knockout mice showed a decreased respiration and increased sensitivity against UVB induced apoptosis [19]. The use of a catalytically inactive mutant TERT did show a smaller effect on respiration and decrease in ROS than wt TERT. This does not necessarily suggest that telomerase activity is important for its mitochondrial function, because the observed differences could as well be caused by other side effects of the mutation, which might for instance impact on mitochondrial uptake of the HA-tagged protein.

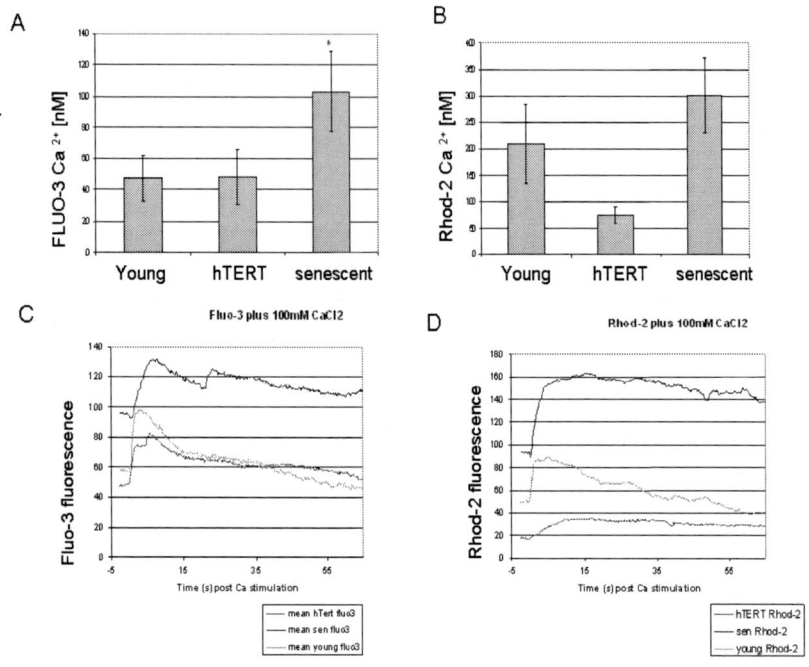

Figure 2. Mitochondrial calcium level is reduced, but calcium storage capacity is increased in hTERT overexpressing cells
A,B, resting calcium level in young, hTERT overexpressing and senescent MRC-5 MRC-5 fibroblasts A; cytoplasmic level measured using Fluo-3 probe, B: mitochondrial level measured using Rhod-2 probe using life cell imaging C, D: Calcium uptake capacity in young (green), hTERT overexpressing (red) and senescent (blue) MRC-5 after stimulation with 100mM $CaCl_2$ using life cell imaging. C: cytoplasmic levels measured using Fluo-3 probe, D: mitochondrial levels measured using Rhod-2 probe (microscopy and data analysis: Glyn Nelson, Newcastle University).

It is not entirely clear whether hTERT resides inside the mitochondrial matrix [19] or rather at the inner membrane (Saretzki, unpublished, Figure 3). Since mitochondrial DNA is known to be closely tethered to the inner mitochondrial membrane, both locations seem plausible to explain the described effects of mitochondrial TERT. In addition, it has been shown for p53 that it can reside in different mitochondrial compartments with opposite functions- either protective or pro-apoptotic [78].

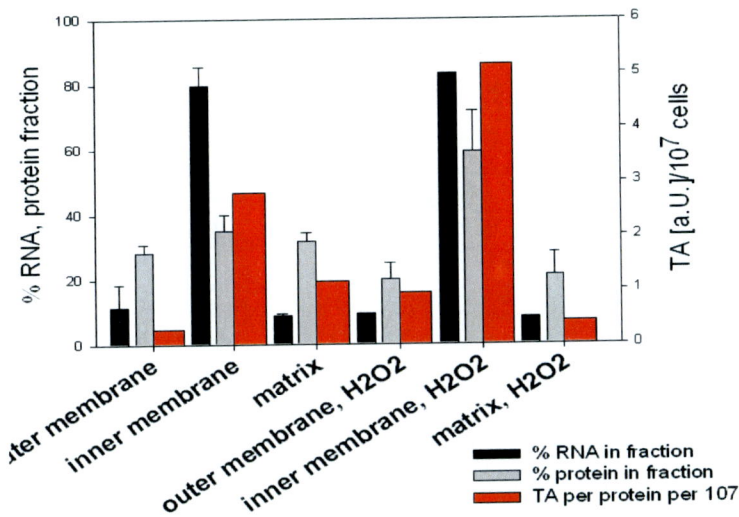

Figure 3. RNA, protein content and telomerase activity in mitochondrial fractions fractions after oxidative stress
Mitochondria were isolatetd from hTERT overexpressing fibroblasts from controls or after 3h 500µM H_2O_2 treatment, treated with RNAse, and protease and then fractionated using digitonin and freeze/thawing cycles. RNA and protein content (Bradford) have been measured in a spectrophotometer. Telomerase activity was determined using TRAP ELISA (Roche). All values were normalysed per 10 million cells. While RNA level does not seem to change after oxidative stress, protein content and telomerase activity increase in the inner membrane fraction after oxidative stress.

Data from Maida et al, [25] suggest that hTERT binds various mitochondrial tRNAs and could therefore interfere with mitochondrial translation.

One so far unresolved area is the localisation of the hTR component with mitochondria. Demonstration of telomerase activity within the mitochondria [15, 19, 20] suggest indirectly that the RNA component of telomerase is present inside the mitochondria as well. RNA transport into mammalian mitochondria is a rare event [80]. However, a combined transport of the RNA component and TERT seems possible but has yet to be proven. In mitochondrial sub fractionation experiments, TERT protein content and telomerase activity in the inner membrane fraction changed in parallel while the RNA content did not change after applied oxidative stress (Figure 3).

7. Telomerase and Oxidative Stress

Two groups recently demonstrated a beneficial role of TERT within mitochondria [19, 20]. Both labs found an improved mitochondrial function, decreased apoptosis and lower oxidative stress in cells expressing telomerase, specifically if directed towards mitochondria. Knockdown experiments with siRNA against telomerase in HUVECs and shRNA HEK293 cells consistently showed an increase in ROS and oxidative stress under these conditions [19, 20].

Interestingly, another recent paper demonstrated that cells and tissues from TERC knockout mice have higher levels of oxidative stress and an imbalance in the cellular redox system. Perez-Rivero and colleagues found in MEFS and tissues (renal cortex) from TERC knockout mice increased MnSOD levels together with decreased catalase accompanied by higher oxidative stress and oxidative damage already in the first generation [64]. The over-expression of MnSOD and the decrease of catalase together can induce an inbalance of antioxidant enzymes that might cause the increased oxidative stress. On the other hand, the increase in MnSOD could be an attempted adaptation to higher ROS in TERC -/- cells. Importantly, this increase in ROS was found in early generation TERC-/- mice with long telomeres and the TERT protein present, and it could be rescued by re-introduction of TERC together with the restoration of telomerase activity. This *in vivo* demonstration of a direct relationship between telomerase deficiency and oxidative stress is in accordance with data from Haendeler et al [19] who showed a decreased oxygen uptake in heart tissue from TERT knockout mice and a better survival of TERT+/+ cells in comparison to MEFs derived from TERT-/- mice. These data are complemented by *in vitro* data demonstrating a decrease of oxidative stress together with higher catalase protein levels and improvement of mitochondrial function in hTERT overexpressing fibroblasts [20, Saretzki unpublished, Figure 4). However, alternative possibilities cannot be ruled out at present. For instance, it is not known whether the knockout of TERC could attenuate mechanisms for the subcellular distribution of TERT. Taken together, the data of Haendeler et al. and Perez-Rivero and colleagues [19, 64] seem to suggest that telomerase activity rather than the TERT protein alone is important for the maintenance of the cellular redox balance.

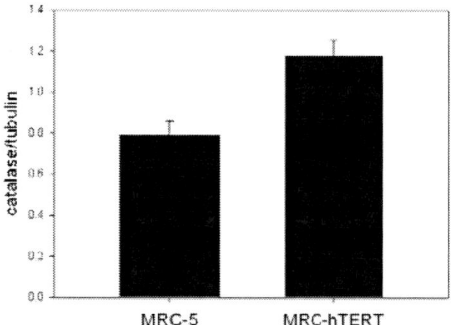

Figure 4. Catalase protein level in parental and hTERT-overexpressing MRC-5 fibroblasts. Western blot analysis was performed and catalase protein levels normalised to tubulin. Data are mean and SE from 4 independent experiments. Courtesy of Ms. Lin Hsiao-Shih

8. Telomerase and Stem Cells

It is well-known that most stem and progenitor cells depend on telomerase activity for maintaining their viability and self-renewal potential [41, 81]. While embryonic stem cells express high levels of telomerase constitutively, adult stem cells are normally dependent on its conditional induction upon activation and proliferation. However, not all functions and mechanisms of telomerase in stem cells are well understood.

It has been demonstrated that in embryonic stem cells telomerase is necessary for continuous proliferation and pluripotency [42, 82]. The high expression of telomerase in these cells correlates to the expression of genes coding for antioxidants and low intracellular oxidative stress [82]. These results correspond very well to the "immortalisation signature" described by Dairkee et al. [49]. This is also in line with the finding that induced pluripotency cells have high levels of telomerase [83, 84] and it is much more difficult to generate induced pluripotency (iPS) cells from late generation telomerase knock-out mice [83].

There is not much known about the potential role of telomerase in adult stem cells. Not all of them even express telomerase [85] and differences in the sensitivity of adult and embryonic stem cells to telomere loss and chromosomal instability have been reported [86]. Telomerase has been shown to improve function and regenerative properties of epidermal stem cells by both telomere dependent [81] and telomere independent mechanisms [11, 41, 87]. Normal tissue stem cells have also been suggested to be responsible to transform into cancer stem cells due to their potential telomerase expression; however, final direct proof for this hypothesis is still missing.

Sarin et al [87] showed that conditional transgenic induction of TERT can activate epidermal skin stem cells causing proliferation of quiescent, multipotent stem cells in the hair follicle bulge region. This developmental transition does not require the catalytic activity of telomerase. These results demonstrated for the first time that TERT has an important telomere-independent function in stem and progenitor cells.

The same group has shown that one way that telomerase can have an effect on development and function of stem cells is by changing gene expression and certain signalling pathways as already described earlier [41]. In an inducible system the authors overexpressed a catalytic inactive TERT that also suppressed endogenous telomerase activity in the epidermal stem cell compartment. This method enabled them to analyse short-term changes in gene expression and they identified an induction of signalling pathways for MYC and WNT by turning off TERT. The authors suggest that these 3 factors could mutually replace each other which could be an explanation why telomerase knock-out mice have a relatively normal embryonic development. Whether this function is restricted to stem cells or even, more specifically, to hair follicle stem cells or can be applied to other cell types as well still has to be shown.

Another follow-up paper from Steve Artandi's group revealed more details of the mechanism how telomerase activates tissue progenitor cells and interferes with the WNT developmental signalling programme [11]. They demonstrate that TERT binds to various promoters of the WNT/β-catenin pathway and directly modulates this pathway by serving as a cofactor in the β-catenin transcriptional complex (See also part 3 "changes in gene expression"). Neither the RNA component nor the catalytic activity of telomerase was necessary for the chromatin changes and the transcriptional activation of the WNT pathway by TERT. Importantly, it seems that TERT and the Wnt/beta-catenin pathways can substitute for each other. This might explain why TERT knockout mice have no major developmental defects in their stem cells despite telomerase being an important player in stem cell biology.

Going one step further, Park et al., show that TERT has an essential role for the correct formation of the anterior-posterior axis in *Xenopus laevis* embryonic development [11]. The Wnt pathway also plays an important role in somite generation from mesoderm in vertebrates including mammals with Wnt3A and Hox genes being important downstream factors. Analysing the first generation of TERT knockout mice, the authors detected homeotic transformations of several vertebrae resulting in the loss of 1 rib in 50% of the G1 pups [11]. Again, these changes are very similar to Wnt3A mutants and demonstrate that there is a phenotypical defect in mice lacking TERT, which might not exist in TERC knockout mice that still retain TERT protein without any catalytic activity.

In addition to direct effects of TERT and the Wnt/beta-catenin pathway on stem cells, the authors also show that it can influence the stem cell niche and therefore has a dual role. Their data identify telomerase as a new transcriptional co-factor that most likely influences stem and progenitor cells in order to facilitate a WNT regulated developmental programme of self renewal, proliferation and perhaps survival.

9. Telomerase and Ageing

The role of telomerase for organismal ageing has been recently addressed in a model of cancer resistant mice [88]. In this model, mice constitutively overexpress TERT in the stem and proliferative compartments of epithelial tissues [89] resulting in a tenfold higher telomerase activity, increased wound healing and improved epidermal stem cell function. In order to diminish the tumourigenic properties of telomerase but to keep its beneficial properties these mice were crossed with cancer resistant mice due to enhanced expression of various tumour suppressors. These transgenic animals displayed a decreased occurrence of degenerative and inflammatory pathologies. They also showed an increased fitness and epithelial barrier function in the GI tract in older age mice, improved glucose tolerance and neuromuscular coordination, less DNA damage and sustained IGF1 levels. In agreement with these features, these mice, if they remained cancer-free, had an increased median life span of up to 50%. Although the authors explain most of the effects by a delay in telomere loss with age that has been demonstrated recently in the stem cell and differentiated compartments [90] it can be speculated that telomere-independent effects of telomerase might play a role as well.

Another example of how telomerase can influence ageing and fitness of certain organs has been demonstrated on heart function in mice and in circulating leukocytes in humans by Werner et al [91, 92]. The authors used a system of voluntary running over 6 months in mice and analysed young and middle-aged endurance athletes. The authors showed that physical exercise can stimulate telomerase activity and the expression of telomere binding proteins in mouse heart and human lymphocytes [91, 92]. Physical exercise has beneficial anti-ageing effects by decreasing senescence associated markers such as p16, p53 and Chk2 and a decrease in lipopolysaccharide-induced aortic endothelial cell apoptosis while the beneficial effect of exercise was absent in TERT negative mice [91, 92]. The age-delaying effects of exercise were also dependent on e-NOS and IGF-1 demonstrating a synergistic effect of these factors and telomerase on endothelial stress resistance after physical activity. This describes a mechanism for effects of exercise for the health of the cardiovascular system. Telomerase levels and telomere length have been shown to decrease in myocytes [93] and other stem cells and differentiated compartments [90] from ageing mice. IGF-1 overexpressing mice also have higher telomerase levels in older age due to persistent AKT-activation. After 3 weeks of voluntary running, TERT expression and telomerase activity roughly doubled in heart tissue. TERT-/- mice did not show an upregulation of TRF2 or decrease of senescence markers. It was known previously that physical exercise increases serum levels of IGF-1 and the authors demonstrated that also the specific levels of IGF-1 in heart tissue increased significantly and could therefore be a mediator of increased TERT levels and telomerase activity. In order to prove causality the authors treated animals with growth hormone and IGF-1 and found a substantial increase of telomerase in both cases [91]. The IGF-1 treatment also resulted in an activation of AKT and e-NOS in the heart. AKT is known to up regulate telomerase activity [94]. Corresponding to better telomere capping and higher telomerase levels there was a higher resistance to doxorubicin induced apoptosis in cardiomyocytes from telomerase wild type, but not knockout mice. The increase in TRF2 and telomerase could either contribute to telomere stabilisation, including the T-loop, or act via telomere-independent mechanisms of telomerase on cardiac stem cells. However, it is not possible to discriminate between these two possibilities from these experiments. The increased proliferation rate in cardiomyocytes could be an attribute of increased telomerase activity and TERT levels interfering with transcriptional activation of certain pathways and growth promoting genes as shown by other studies [8, 41, 87]. For TRF2 telomere independent functions have been described as well [95]. While the authors failed to detect a change in telomere length in mice during 6 month of exercise they were able to demonstrate a decrease in telomere shortening in athletes with long-term endurance training compared with untrained controls. Since it is known that oxidative stress can modulate telomere length as well as senescence markers such as p53, Chk2 and p16, it is tempting to speculate that telomerase upregulation could cause the decrease of oxidative stress as shown before [19, 20].

10. TERC-Independent Biochemical Activities of TERT

Polymerases are dependent on certain metals for their activity. Usually these are divalent metal ions. In 2005, Lue and colleagues described that yeast and human telomerase switch their biochemical activities in the presence of manganese to being a terminal transferase [24]. The authors showed that in the presence of Mn2+ ions TERT can act in a template independent fashion. It could be speculated that several non-telomeric functions, for example, the effects of over expression of TERT on the induction of senescence and tumourigensis [7, 96] could be explained by other RNA independent biochemical activities of telomerase that become unmasked under conditions of unbalanced expression of both components. However, it is not known yet whether the *in vitro* conditions of Mn^{2+} ion concentrations can be reached *in vivo*. However, it is possible that other factors than Mn^{2+} ions can initiate this alternative function for TERT.

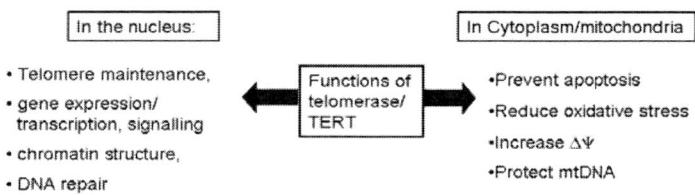

Figure 5. Known functions of telomerase/TERT in different cellular compartments.

Recently, Maida et al. have found another example for the remarkable plasticity of telomerase [25]. By switching the RNA component from TERC to the RNA component of MRP (Mitochondrial RNAseP) the enzyme switches from being an RNA dependent DNA polymerase to being an RNA dependent RNA polymerase. This new enzyme generates double stranded RNAs that can be processed into small interfering RNAs. These siRNAs regulate the amount of the RNA component of the mitochondrial RNA processing endoribonuclease (RMRP). Despite the misleading name (the enzyme was initially described in mitochondria) RMRP is a small, non-coding RNA that can also be found in mitochondria. In the nucleus it is involved in the processing of ribosomal RNAs. Importantly, the authors found that the level of TERT in a cell determines the level of RMRP via a negative feedback-loop: introduction of RMRP into cells without TERT increases cellular RMRP levels. In contrast, the over expression of RMRP in cells that do express TERT actually decreases cellular RMRP levels. Consistently, repression of TERT in Hela cells increased, but overexpression of TERT and RMRP in telomerase negative cells decreased the levels of RMRP. The authors suggest that endogenously encoded siRNAs could have a role in the regulation of gene expression. Given its function in the regulation of ribosomal RNAs, it is tempting to speculate that this internal process regulates the biogenesis of ribosomes and thereby the overall level of protein synthesis of a cell.

In addition to RMRP that complexes with TERT with the same stochiometry as the telomerase TERC component, the authors identified several more RNAs that had been found in a complex with TERT, including a large number of mitochondrially encoded tRNAs. It will be interesting to determine whether these complexes actually exist in cells and contribute to a non-telomeric mitochondrial function of hTERT. One possibility would be the interference with mitochondrial translation. Known non-telomeric functions of telomerase are summarised in figure 5.

Conclusion

In addition to the classical, canonical role of telomerase in telomere maintenance, various non-canonical roles have been described. Many of those telomere-independent functions seem not to require catalytic activity or the involvement of the RNA component TR. Non-canonical functions of telomerase reach from TERT being a transcriptional co-factor that regulates several signalling pathways during development of stem cells to mitochondrial localisation of telomerase and even to influencing behaviour such as anxiety. Not all these functions or their mechanisms are well understood and might co-exist within a cell with telomere-dependent ones. This sometime confuses matters and creates controversies. This applies in particular to functions of TERT that are exerted within the same compartment as telomere maintenance: the nucleus. With the help of mutant TERT proteins without catalytic function it was already possible to dissect several important non-telomeric functions such as changes in gene expression, transcriptional activation and interference with developmental signalling pathways. Other useful approaches could be the detailed comparison of genetic knock-out models for the two main telomerase components and the specific expression of telomerase in certain cellular locations such as the nucleus and mitochondria. This could, for example, help to analyse in which cellular compartment TERT has to reside for its anti-apoptotic function. Due to the central role of mitochondria for intracellular oxidative stress, signalling, metabolism and apoptosis, it seems imperative to further analyse the mechanism of interference of mitochondrial telomerase with these important organellar and cellular functions.

References

[1] Greider, CW; Blackburn, EH. Identification of a specific telomere terminal transferase activity in Tetrahymena extracts. *Cell*, 1985, Dec, 43(2 Pt 1), 405-13.
[2] Fossel, M. *Reversing human aging*. 1st Edition, New York, 1996, William Morrow & Co.
[3] Nakamura, TM; Morin, GB; Chapman, KB; Weinrich, SL; Andrews, WH; Lingner, J; Harley, CB; Cech, TR. Telomerase catalytic subunit homologs from fission yeast and human. *Science*, 1997, Aug 15, 277(5328), 955-9.
[4] Leeper, T; Leulliot, N; Varani, G. The solution structure of an essential stem-loop of human telomerase RNA. Nucleic Acids Res., 2003, 15, 31(10), 2614-21.
[5] Wright, WE; Piatyszek, MA; Rainey, WE; Byrd, W; Shay, JW. Telomerase activity in human germline and embryonic tissues and cells. Dev Genet. 1996, 18(2), 173-9.
[6] *Shay, JW; Bacchetti, S*. A survey of telomerase activity in human cancer. Eur J Cancer. 1997, 33(5), 787-91.
[7] *Stewart, SA; Hahn, WC; O'Connor, BF; Banner, EN; Lundberg, AS; Modha, P; Mizuno, H; Brooks, MW; Fleming, M; Zimonjic, DB; Popescu, NC; Weinberg, RA*. Telomerase contributes to tumorigenesis by a telomere length-independent mechanism. Proc Natl Acad Sci U S A. 2002, Oct 1, 99(20), 12606-11.
[8] Smith, LL; Coller, HA; Roberts, JM; Telomerase modulates expression of growth-controlling genes and enhances cell proliferation. *Nat. Cell, Biol*, 2003, 5, 474-479.

[9] Li, S; Crothers, J; Haqq, CM; Blackburn, EH; Cellular and gene expression responses involved in the rapid growth inhibition of human cancer cells by RNA interference-mediated depletion of telomerase RNA. *J. Biol. Chem.*, 2005, 280, 23709-23717.

[10] Bagheri, S; Nosrati, M; Li, S; Fong, S; Torabian, S; Rangel, J; Moore, DH; Federman, S; Laposa, RR; Baehner, FL; Sagebiel, RW; Cleaver, JE; Haqq, C; Debs, RJ; Blackburn, EH; Kashani-Sabet, M. Genes and pathways downstream of telomerase in melanoma metastasis. Proc Natl Acad Sci U S A. 2006, Jul 25, 103(30), 11306-11.

[11] Park, JI; Venteicher, AS; Hong, JY; Choi, J; Jun, S; Shkreli, M; Chang, W; Meng, Z; Cheung, P; Ji, H; McLaughlin, M; Veenstra, TD; Nusse, R; McCrea, PD; Artandi, SE. Telomerase modulates Wnt signalling by association with target gene chromatin. Nature. 2009, Jul 2, 460(7251), 66-72.

[12] Zhang, P; Chan, SL; Fu, W. Mendoza M, Mattson MP TERT suppresses apoptosis at a premitochondrial step by a mechanism requiring reverse transcriptase activity and 14-3-3 protein-binding ability. *FASEB J*, 2003, 17, 767-769.

[13] Massard, C; Zermati, Y; Pauleau, AL; Larochette, N; Metivier, D; Sabatier, L; Kroemer, G; Soria, JC. hTERT, a novel endogenous inhibitor of the mitochondrial cell death pathway. *Oncogene*, 2006, 25, 4505-4514.

[14] Del Bufalo, D; Rizzo, A; Trisciuoglio, D; Cardinali, G; Torrisi, MR; Zangemeister-Wittke, U; Zupi, G; Biroccio, A. Involvement of hTERT in apoptosis induced by interference with Bcl-2 expression and function. *Cell Death Differ*, 2005, 12, 1429-1438.

[15] Santos, JH; Meyer, JN; Skorvaga, M; Annab, LA; Van Houten, B. Mitochondrial hTERT exacerbates free-radical-mediated mtDNA damage. *Aging Cell*, 2004, 3, 399-411.

[16] Santos, JH; Meyer, JN; Van Houten, B. Mitochondrial localization of telomerase as a determinant for hydrogen peroxide-induced mitochondrial DNA damage and apoptosis. *Hum Mol Genet*, 2006, 15(11), 1757-68.

[17] Haendeler, J; Hoffmann, J; Brandes, RP; Zeiher, AM; Dimmeler, S. Hydrogen peroxide triggers nuclear export of telomerase reverse transcriptase via Src kinase family-dependent phosphorylation of tyrosine 707. *Mol Cell Biol.*, 2003, 23(13), 4598-610.

[18] Haendeler, J; Hoffmann, J; Diehl, JF; Vasa, M; Spyridopoulos, I; Zeiher, AM; Dimmeler, S. Antioxidants inhibit nuclear export of telomerase

reverse transcriptase and delay replicative senescence of endothelial cells. *Circ Res.*, 2004, 94(6), 768-75.
[19] Haendeler, J; Dröse, S; Büchner, N; Jakob, S; Altschmied, J; Goy, C; Spyridopoulos, I; Zeiher, AM; Brandt, U; Dimmeler, S. Mitochondrial telomerase reverse transcriptase binds to and protects mitochondrial DNA and function from damage. *Arterioscler Thromb Vasc Biol.*, 2009, Jun, 29(6), 929-35.
[20] Ahmed, S; Passos, JF; Birket, MJ; Beckmann, T; Brings, S; Peters, H; Birch-Machin, MA; von Zglinicki, T; Saretzki, G. Telomerase does not counteract telomere shortening but protects mitochondrial function under oxidative stress. *J. Cell Science*, 2008, 17, 1046-1053.
[21] Kang HJ; Choi YS; Hong, SB; et al Ectopic expression of the catalytic subunit of telomerase protects against brain injury resulting from ischemia and NMDA-induced neurotoxicity. *J Neurosci*, 2004, 24, 1280-1287.
[22] Indran, IR; Pervaiz, S. Tumor cell redox state and mitochondria at the center of the non-canonical activity of telomerase reverse transcriptase. Mol Aspects Med. 2009 Dec 6. [Epub ahead of print]
[23] Lee, J; Jo, YS; Sung, YH; Hwang, IK; Kim, H; Kim, SY; Yi, SS; Choi, JS; Sun, W; Seong, JK; Lee, HW. Telomerase Deficiency Affects Normal Brain Functions in Mice. Neurochem Res. 2009 Aug 15. Epub ahead of print.
[24] Lue, NF; Bosoy, D; Moriarty, TJ; Autexier, C; Altman, B; Leng, S. Telomerase can act as a template- and RNA-independent terminal transferase. Proc Natl Acad Sci U S A. 2005, 102(28), 9778-83.
[25] Maida, Y; Yasukawa, M; Furuuchi, M; Lassmann, T; Possemato, R; Okamoto, N; Kasim, V; Hayashizaki, Y; Hahn, WC; Masutomi, K. An RNA-dependent RNA polymerase formed by TERT and the RMRP RNA. *Nature*, 2009 Sep 10, 461(7261), 230-5.
[26] McClintock, B. The Behavior in Successive Nuclear Divisions of a Chromosome Broken at Meiosis. Proc Natl Acad Sci U S A. 1939, Aug, 25(8), 405-16.
[27] Heller, K; Kilian, A; Piatyszek, MA; Kleinhofs, A. Telomerase activity in plant extracts. *Mol Gen Genet*, 1996, 252(3), 342-5.
[28] Müller, F; Wicky, C; Spicher, A; Tobler, H. New telomere formation after developmentally regulated chromosomal breakage during the process of chromatin diminution in Ascaris lumbricoides. Cell. 1991, 67(4), 815-22.

[29] Bottius, E; Bakhsis, N; Scherf, A. Plasmodium falciparum telomerase: de novo telomere addition to telomeric and nontelomeric sequences and role in chromosome healing. *Mol Cell Biol.*, 1998, 18(2), 919-25.
[30] Sandell, LL; Zakian, VA. Loss of a yeast telomere: arrest, recovery, and chromosome loss. *Cell*, 1993, 75, 729-739.
[31] *Wilkie, AO; Lamb, J; Harris, PC; Finney, RD; Higgs, DR.* A truncated human chromosome 16 associated with alpha thalassaemia is stabilized by addition of telomeric repeat (TTAGGG)n. Nature. 1990, Aug 30, 346(6287), 868-71.
[32] *Flint, J; Craddock, CF; Villegas, A; Bentley, DP; Williams, HJ; Galanello, R; Cao, A; Wood, WG; Ayyub, H; Higgs, DR.* Healing of broken human chromosomes by the addition of telomeric repeats. Am J Hum Genet. 1994, Sep, 55(3), 505-12.
[33] *Varley, H; Di, S; Scherer, SW; Royle, NJ.* Characterization of terminal deletions at 7q32 and 22q13.3 healed by De novo telomere addition. Am J Hum Genet. 2000, Sep, 67(3), 610-22.
[34] Morin, GB. Recognition of a chromosome truncation site associated with alpha-thalassaemia by human telomerase. *Nature*, 1991, Oct 3, 353(6343), 454-6.
[35] *Fitzgerald, MS; Shakirov, EV; Hood, EE; McKnight, TD; Shippen, DE.* Different modes of de novo telomere formation by plant telomerases. Plant J. 2001, 26(1), 77-87.
[36] Schul,z, VP; Zakian, VA. The saccharomyces PIF1 DNA helicase inhibits telomere elongation and de novo telomere formation. *Cell*, 1994, 76, 145-155.
[37] *Makovets, S; Blackburn, EH.* DNA damage signalling prevents deleterious telomere addition at DNA breaks. Nat Cell Biol. 2009, Nov, 11(11), 1383-6.
[38] Xiang, H; Wang, J; Mao, Y; Liu, M; Reddy, VN; Li, DW. Human telomerase accelerates growth of lens epithelial cells through regulation of the genes mediating RB/E2F pathway. *Oncogene*, 2002, May 23, 21(23), 3784-91.
[39] Dairkee, SH; Nicolau, M; Sayeed, A; Champion, S; Ji, Y; Moore, DH; Yong, B; Meng, Z; Jeffrey, SS. Oxidative stress pathways highlighted in tumor cell immortalization, association with breast cancer outcome. *Oncogene*, 2007, Sep 20, 26(43), 6269-79.
[40] Wang, J; Feng, H; Huang, XQ; Xiang, H; Mao, YW; Liu, JP; Yan, Q; Liu, WB; Liu, Y; Deng, M; Gong, L; Sun, S; Luo, C; Liu, SJ; Zhang, XJ; Liu, Y; Li, DW. Human telomerase reverse transcriptase

immortalizes bovine lens epithelial cells and suppresses differentiation through regulation of the ERK signaling pathway. *J Biol Chem.*, 2005, 280(24), 22776-87.

[41] Choi, J; Southworth, LK; Sarin, KY; Venteicher, A.S; Ma, W; Chang, W; Cheung, P; Jun, S; Artandi, MK; Shah, N; Kim, SK; Artandi, SE. TERT promotes epithelial proliferation through transcriptional control of a Myc- and Wnt-related developmental program. *PLoS Genet*, 2008, 1, e10.

[42] Yang, C; Pryzborski, S; Cooke, MJ; Zhang, X; Stewart, R; Atkinson, S; Saretzki, G; Armstrong, L; Lako, M. A key role for telomerase reverse transcriptase unit (TERT) in modulating human ESC proliferation, cell cycle dynamics and *in vitro* differentiation. *Stem Cells*, 2008, 26, 850-863.

[43] Masutomi, K; Possemato, R; Wong, JM; Currier, JL; Tothova, Z; Manola, JB; Ganesan, S; Lansdorp, PM; Collins, K; Hahn, WC. The telomerase reverse transcriptase regulates chromatin state and DNA damage responses. *Proc. Natl. Acad. Sci. U. S. A* , 2005, 102, 8222-8227

[44] Sharma, GG; Gupta, A; Wang, H; Scherthan, H; Dhar, S; Gandhi, V; Iliakis, G; Shay, JW; Young, CS; Pandita, TK. hTERT assocates with human telomeres and enhances genomic stability and DNA repair. *Oncogene*, 2003, 22, 131-146.

[45] Masutomi, K; Yu, EY; Khurts, S; Ben-Porath, I; Currier, JL; Metz, GB; Brooks, MW; Kaneko, S; Murakami, S; DeCaprio, JA; Weinberg, RA; Stewart, SA; Hahn, WC. Telomerase maintains telomere structure in normal human cells. *Cell*, 2003, Jul 25, 114(2), 241-53.

[46] Bradshaw, PS; Stavropoulos, DJ; Meyn, MS. Human telomeric protein TRF2 associates with genomic double-strand breaks as an early response to DNA damage. *Nat Genet*, 2005, Feb, 37(2), 193-7.

[47] *Wong, JM; Kusdra, L; Collins, K.* Subnuclear shuttling of human telomerase induced by transformation and DNA damage. Nat Cell Biol. 2002, Sep, 4(9), 731-6.

[48] Zhu, H; Fu, W; Mattson, MP; The catalytic subunit of telomerase protects neurons against amyloid beta-peptide-induced apoptosis. *J. Neurochem*, 2000, 75, 117-124.

[49] Lu, C; Fu, W; Mattson, MP. Telomerase protects developing neurons against DNA damage-induced cell death. *Brain Res Dev Brain Res.*, 2001, Nov 26, 131(1-2), 167-71.

[50] Fu, W; Killen, M; Culmsee, C; Dhar, S; Pandita, TK; Mattson, MP. The catalytic subunit of telomerase is expressed in developing brain neurons

and serves a cell survival-promoting function. *J. Mol. Neurosci*, 2000, 14, 3-15.
[51] Lee, J; Sung, YH; Cheong, C; et al TERT promotes cellular and organismal survival independently of telomerase activity. *Oncogene*, 2008, 27, 3754-3760.
[52] Gorbunova, V; Seluanov, A; Pereira-Smith, OM. Expression of human telomerase (hTERT) does not prevent stress-induced senescence in normal human fibroblasts but protects the cells from stress-induced apoptosis and necrosis. *J Biol Chem.*, 2002, 277(41), 38540-9.
[53] Kondo, Y; Kondo, S; Tanaka, Y; Haqqi, T; Barna, BP; Cowell, JK. Inhibition of telomerase increases the susceptibility of human malignant glioblastoma cells to cisplatin-induced apoptosis. *Oncogene*, 1998, 16, 2243-2248.
[54] Kondo, Y; Komata, T; Kondo, S.Combination therapy of 2-5A antisense against telomerase RNA and cisplatin for malignant gliomas. *Int J Oncol.*, 2001, Jun, 18(6), 1287-92.
[55] Ludwig, A; Saretzki, G; Holm, PS; Tiemann, F; Lorenz, M; Emrich, T; Harley, CB; von Zglinicki, T. Ribozyme cleavage of telomerase mRNA sensitizes breast epithelial cells to inhibitors of topoisomerase. *Cancer Res.*, 2001, 61, 3053-3061.
[56] Saretzki, G; Ludwig, A; von Zglinicki, T; Runnebaum, IB. Ribozyme-mediated telomerase inhibition induces immediate cell loss but not telomere shortening in ovarian cancer cells. *Cancer Gene Ther.*, 2001, 8, 827-834.
[57] *Folini, M; Brambilla, C; Villa, R; Gandellini, P; Vignati, S; Paduano, F; Daidone, MG; Zaffaroni, N.* Antisense oligonucleotide-mediated inhibition of hTERT, but not hTERC, induces rapid cell growth decline and apoptosis in the absence of telomere shortening in human prostate cancer cells. Eur J Cancer. 2005, Mar, 41(4), 624-34.
[58] Rubio, MA; Davalos, AR; Campisi, J. Telomere length mediates the effects of telomerase on the cellular response to genotoxic stress. *Exp Cell Res.*, 2004, 298(1), 17-27.
[59] Counter, CM; Hahn, WC; Wei, W; Caddle, SD; Beijersbergen, RL; Lansdorp, PM; Sedivy, JM; Weinberg, RA. Dissociation among in vitro telomerase activity, telomere maintenance, and cellular immortalization. *Proc Natl Acad Sci U S A*, 1998, Dec 8, 95(25), 14723-8.
[60] Luiten, RM; Péne, J; Yssel, H; Spits, H. Ectopic hTERT expression extends the life span of human CD4+ helper and regulatory T-cell clones

and confers resistance to oxidative stress-induced apoptosis. *Blood*, 2003, Jun 1, 101(11), 4512-9.
[61] Cao, Y; Li, H; Deb, S; Liu, JP; TERT regulates cell survival independent of telomerase enzymatic activity. *Oncogene*, 2002, 21(20), 3130-8.
[62] Rahman, R; Latonen, L; Wiman, KG. hTERT antagonizes p53-induced apoptosis independently of telomerase activity. *Oncogene*, 2005, Feb 17, 24(8), 1320-7.
[63] Parkinson, EK; Fitchett, C; Cereser, B. Dissecting the non-canonical functions of telomerase. *Cytogenet Genome Res.*, 2008, 122(3-4), 273-80.
[64] Pérez-Rivero, G; Ruiz-Torres, MP; Díez-Marqués, ML; Canela, A; López-Novoa, JM; Rodríguez-Puyol, M; Blasco, MA; Rodríguez-Puyol, D. Telomerase deficiency promotes oxidative stress by reducing catalase activity. *Free Radic Biol Med*, 2008, 45(9), 1243-51.
[65] Boklan, J; Nanjangud, G; MacKenzie, KL; May, C; Sadelain, M; Moore, MA. Limited proliferation and telomere dysfunction following telomerase inhibition in immortal murine fibroblasts. Cancer Res. 2002, Apr 1, 62(7), 2104-14.
[66] Klapper, W; Shin, T; Mattson, MP. Differential regulation of telomerase activity and TERT expression during brain development in mice. *J Neurosci Res.*, 2001, 64, 252-260.
[67] *Caporaso, GL; Lim, DA; Alvarez-Buylla, A; Chao, MV.* Telomerase activity in the subventricular zone of adult mice. Mol Cell Neurosci. 2003, Aug, 23(4), 693-702.
[68] Lee, J; Jo, YS; Sung, YH; Hwang, IK; Kim, H; Kim, SY; Yi, SS; Choi, JS; Sun, W; Seong, JK; Lee, HW. Telomerase Deficiency Affects Normal Brain Functions in Mice. *Neurochem Res.*, 2009, Aug 15. [Epub ahead of print]
[69] *Kim, WR; Kim, Y; Eun, B; Park, OH; Kim, H; Kim, K; Park, CH; Vinsant, S; Oppenheim, RW; Sun, W.* Impaired migration in the rostral migratory stream but spared olfactory function after the elimination of programmed cell death in Bax knock-out mice. J Neurosci. 2007, Dec 26, 27(52), 14392-403.
[70] Seimiya, H; Sawada, H; Muramatsu, Y; Shimizu, M; Ohko, K; Yamane, K; Tsuruo, T. Involvement of 14-3-3 proteins in nuclear localization of telomerase. *EMBO J*, 2000, 19(11), 2652-61.
[71] Teixeira, MT; Forstemann, K; Gasser, SM; Lingner, J. Intracellular trafficking of yeast telomerase components. *EMBO Rep*, 2002, Jul, 3(7), 652-9.

[72] Gallardo, F; Olivier, C; Dandjinou, AT; Wellinger, RJ; Chartrand, P. TLC1 RNA nucleo-cytoplasmic trafficking links telomerase biogenesis to its recruitment to telomeres. *EMBO J*, 2008, Mar 5, 27(5), 748-57.
[73] Wong, JM; Kusdra, L; Collins, K. Subnuclear shuttling of human telomerase induced by transformation and DNA damage. *Nat Cell Biol.*, 2002 Sep, 4(9), 731-6.
[74] Chang, LY; Lin, SC; Chang, CS; Wong, YK; Hu, YC; Chang, KW. Telomerase activity and in situ telomerase RNA expression in oral carcinogenesis. *J Oral Pathol Med*, 1999, Oct, 28(9), 389-96.
[75] Liu, K; Hodes, RJ; Weng, Np. Cutting edge: telomerase activation in human T lymphocytes does not require increase in telomerase reverse transcriptase (hTERT) protein but is associated with hTERT phosphorylation and nuclear translocation. *J Immunol*, 2001, 166(8), 4826-30.
[76] Passos, JF; Nelson, G; Wang, C; Richter, T; Simillion, C. Proctor, CJ; Miwa, S; Olijslagers, S; Hallinan, J; Wipat, A; Saretzki, G; Rudolph, KL; Kirkwood TBL von Zglinicki, T. Feedback between p21 and reactive oxygen production is necessary for cell senescence. *Molecular Systems Biology*, 2010, in press.
[77] Buechner, N; Altschmied, J; Jakob, S; Saretzki, G; Haendeler, J. Well-known signalling proteins exert new functions in the nucleus and mitochondria. *Antioxid Redox Signal*, 2009 Dec 3. [Epub ahead of print]
[78] Vaseva, AV; Moll, UM. The mitochondrial p53 pathway. *Biochim Biophys Acta*, 2009, May, 1787(5), 414-20.
[79] Arachiche, A; Augereau, O; Decossas, M; Pertuiset, C; Gontier, E; Letellier, T; Dachary-Prigent, J. Localization of PTP-1B, SHP-2, and Src exclusively in ratbrain mitochondria and functional consequences. *J Biol Chem.*, 2008, 283, 24406-24411.
[80] Entelis, N; Kolesnikova, O; Kazakova, H; Brandina, I; Kamenski, P; Martin, RP; Tarassov, I. Import of nuclear encoded RNAs into yeast and human mitochondria: experimental approaches and possible biomedical applications. *Genet Eng* (N Y), 2002, 24, 191-213.
[81] Flores, I; Cayuela, ML; Blasco, MA. Effects of telomerase and telomere length on epidermal stem cell behavior. *Science*, 2005, 309, 1253-1256.
[82] Saretzki, G; Walter, T; Atkinson, S; Passos, JF; Bareth, B; Keith, WN; Stewart, R; Hoare, S; Stojkovic, M; Armstrong, L; von Zglinicki, T; Lako, M. Down-regulation of multiple stress defence mechanisms during differentiation of human embryonic stem cells. *Stem Cells*, 2008, 26, 455-465.

[83] Marion, RM; Strati, K; Li, H; Tejera, A; Schoeftner, S; Ortega, S; Serrano, M; Blasco, MA. Telomeres acquire embryonic stem cell characteristics in induced pluripotent stem cells. *Cell Stem Cell*, 2009, 4(2), 141-54.

[84] Armstrong, L; Tilgner, K; Saretzki, G; Atkinson, SP; Stojkovic, M; Moreno, R; Pryzborski, S; Lako, M. Human induced pluripotent stem cell line show similar stress defense mechanisms and mitochondrial regulation to human embryonic stem cells, *Stem Cells*, 2010, in press.

[85] Hiyama, E; Hiyama, K. Telomere and telomerase in stem cells. *Br J Cancer*, 2007, Apr 10, 96(7), 1020-4.

[86] Ferrón, S; Mira, H; Franco, S; Cano-Jaimez, M; Bellmunt, E; Ramírez, C; Fariñas, I; Blasco, MA. Telomere shortening and chromosomal instability abrogates proliferation of adult but not embryonic neural stem cells. Development. 2004, Aug, 131(16), 4059-70.

[87] Sarin, KY; Cheung, P; Gilison, D; Lee, E; Tennen, RI; Wang, E; Artandi, MK; Oro, AE; Artandi, SE. Conditional telomerase induction causes proliferation of hair follicle stem cells. *Nature*, 2005, 436, 1048-1052.

[88] Tomás-Loba, A; Flores, I; Fernández-Marcos, PJ; Cayuela, ML; Maraver, A; Tejera, A; Borrás, C; Matheu, A; Klatt, P; Flores, JM; Viña, J; Serrano, M; Blasco, MA. Telomerase reverse transcriptase delays aging in cancer-resistant mice. *Cell*, 2008, 135(4), 609-22.

[89] González-Suárez, E; Samper, E; Ramírez, A; Flores, JM; Martín-Caballero, J; Jorcano, JL; Blasco, MA. Increased epidermal tumors and increased skin wound healing in transgenic mice overexpressing the catalytic subunit of telomerase, mTERT, in basal keratinocytes. *EMBO J*, 2001, Jun 1, 20(11), 2619-30.

[90] Flores, I; Canela, A; Vera, E; Tejera, A; Cotsarelis, G; Blasco, MA. The longest telomeres: a general signature of adult stem cell compartments. *Genes Dev*, 2008, 1, 22(5), 654-67.

[91] Werner, C; Hanhoun, M; Widmann, T; Kazakov, A; Semenov, A; Pöss, J; Bauersachs, J; Thum, T; Pfreundschuh, M; Müller, P; Haendeler, J; Böhm, M; Laufs, U. Effects of physical exercise on myocardial telomere-regulating proteins, survival pathways, and apoptosis. *J Am Coll Cardiol*, 2008, Aug 5, 52(6), 470-82.

[92] Werner, C; Fürster, T; Widmann, T; Pöss, J; Roggia, C; Hanhoun, M; Scharhag, J; Büchner, N; Meyer, T; Kindermann, W; Haendeler, J; Böhm, M; Laufs, U. Physical exercise prevents cellular senescence in

circulating leukocytes and in the vessel wall. *Circulation*, 2009, Dec 15, 120(24), 2438-47.
[93] Torella, D; Rota, M; Nurzynska, D; Musso, E; Monsen, A; Shiraishi, I; Zias, E; Walsh, K; Rosenzweig, A; Sussman, MA; Urbanek, K; Nadal-Ginard, B; Kajstura, J; Anversa, P; Leri, A. Cardiac stem cell and myocyte aging, heart failure, and insulin-like growth factor-1 overexpression. *Circ Res.*, 2004, Mar 5, 94(4), 514-24.
[94] Kang, SS; Kwon, T; Kwon, DY; Do, SI. Akt protein kinase enhances human telomerase activity through phosphorylation of telomerase reverse transcriptase subunit. *J Biol Chem.*, 1999, 274(19), 13085-90.
[95] *Spyridopoulos, I; Haendeler, J; Urbich, C; Brummendorf, TH; Oh, H; Schneider, MD; Zeiher, AM; Dimmeler, S.* Statins enhance migratory capacity by upregulation of the telomere repeat-binding factor TRF2 in endothelial progenitor cells. Circulation. 2004, Nov 9, 110(19), 3136-42.
[96] Gorbunova, V; Seluanov, A; Pereira-Smith, OM. Evidence that high telomerase activity may induce a senescent-like growth arrest in human fibroblasts. *J Biol Chem.*, 2003, Feb 28, 278(9), 7692-8.

Chapter II

Telomerase and Telomeric Proteins: A Life Beyond Telomeres

Ilaria Chiodi, Cristina Belgiovine
*and Chiara Mondello**
Istituto di Genetica Molecolare, CNR, via Abbiategrasso 207,
27100 Pavia, Italy

Abstract

Telomerase was discovered as the enzyme deputed to telomere maintenance. In human somatic cells, its activity is virtually absent, and cells stop proliferating when telomeres fall below a critical length. In contrast, in the vast majority of tumors, telomerase is active and its activity is associated with telomere preservation and with the acquisition of an unlimited proliferative potential. Despite this well known and consolidated role of telomerase, a large body of evidence indicates that it contributes to cellular physiology independently of telomere length maintenance. Multiple roles of telomerase in cancer development have been highlighted, besides telomere lengthening, among which is gene expression regulation. More recently, the telomerase catalytic subunit has

---

* Corresponding author: E-mail: mondello@igm.cnr.it Phone: +39 0382 546332. Fax: +39 0382 422286.

been shown to associate with an RNA molecule different from the telomerase RNA, giving rise to a complex with functions unrelated to telomere metabolism. Besides telomerase, telomeric proteins play a fundamental role in telomere maintenance. In particular, a six protein complex, named shelterin, is stably associated with telomeres and is essential for telomere metabolism. As well as for telomerase, additional roles beyond telomeres have been found for telomeric proteins, in particular for the TRF2 subunit of shelterin. In this review, we will discuss extra-telomeric functions of telomerase and TRF2 in different biological processes.

Introduction

In the 1930s, Muller coined the term "telomere" (τελος-μερος: final-part) to name chromosome ends, when the observation that chromosomes devoid of their ends were unstable made him hypothesize that this part of the chromosomes had distinctive features compared to the rest of the genome (Muller, 1938). At the beginning of the 1980s, the molecular nature of the telomeric DNA was discovered: in most eukaryotes it is composed of short repetitive G-rich sequences; in most vertebrates, including humans, the basic repeat is the hexamer TTAGGG (Szostak & Blackburn, 1982; Shampay et al., 1984). In mammals, the extension of the telomeric DNA is kept within a species specific range, but it varies among chromosomes of the same cell population, thus it is generally possible to identify an "average telomere length" for different organisms, rather than a defined one (e.g., in humans the average telomere length is around 6–20 kb, depending on age, tissue, etc; in mice around 50 kb). A peculiar feature of the telomeric DNA is the presence of a single strand region at the 3' end. The end of the telomere folds back leading to the formation of a loop (t-loop: telomeric loop), in which, the 3' single strand region invades the double strand region, giving rise to the so-called displacement loop (D-loop; Figure 1A) (Griffith et al., 1999; Blackburn, 2005). Telomeric DNA is sufficient to stabilize linear DNA molecules, but it does not work alone: it interacts with specific proteins (de Lange, 2005) and is replicated by the specialized enzyme telomerase (Greider &Blackburn, 1985).

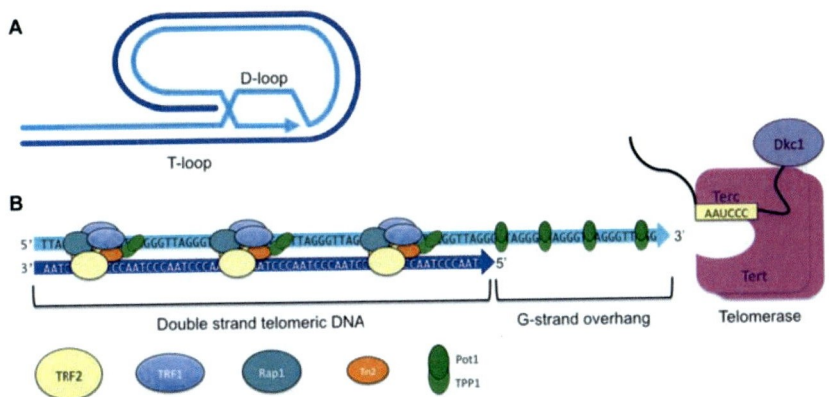

Figure 1. Mammalian telomeres. A) Schematic representation of putative telomere T-loop and D-loop structures. The single-stranded DNA at the end of the telomere is able to invade and anneal with part of the duplex DNA in the same telomere. B) Mammalian telomeres consist of tandem repeats of the double-stranded TTAGGG sequence with a 3' prime single strand overhang. Double strand telomeric repeats are bound by a multiprotein complex known as 'shelterin'. The POT1–TPP1 heterodimer is associated with shelterin, but it binds the G-strand overhang. Telomerase is able to recognize the 3' end of the G-strand overhang to elongate telomeres (Dkc1: diskerin subunit).

In mammals, telomeric DNA is bound to a complex of six proteins named shelterin (de Lange, 2005; Palm & de Lange, 2008). Three of them, TRF1, TRF2 and POT1, bind directly to the telomeric DNA, TRF1 and TRF2 to the double stranded portion, POT1 to the single stranded overhang. TIN2 and TPP1 connect together these three proteins, while RAP1 binds to TRF2 (Figure 1B). The main functions of telomeric proteins are to regulate telomere length and to avoid having the telomeres, being formally equivalent to broken DNA ends, become engaged in pathways involved in DNA damage response or in DNA double strand break repair (de Lange, 2009). Proteins of the shelterin complex interact with several other proteins known to play a role in different cellular processes, mainly DNA damage response and DNA repair.

Telomerase adds telomeric repeats to the 3' end of the telomeric DNA, solving the "end replication problem", due to the inability of DNA polymerases to complete the replication of 3' DNA ends. Catalytically active telomerase is composed of three main subunits: the catalytic subunit, with reverse transcriptase activity (TERT: telomerase reverse transcriptase), an RNA molecule, containing the template for the reverse transcriptase (TR:

telomeric RNA), and dyskerin (Greider & Blackburn, 1985; Cohen et al., 2007; Collins, 2008). In human cells, expression of hTERT (human TERT) is the limiting factor for telomerase activity (Bodnar et al., 1998). Therefore, in somatic cells, in which hTERT is virtually absent (Masutomi et al., 2003), telomerase activity is not sufficient to maintain telomeres, which shorten at each cell division (Harley et al., 1990; Harley, 1991). Telomere maintenance and cellular proliferation are two strictly connected processes. In fact, when telomeres shorten below a threshold level, they trigger a proliferation arrest, well known as cellular senescence (Campisi & d'Adda di Fagagna, 2007); in contrast, cells which are able to maintain telomeres, can proliferate indefinitely. A paradigmatic example of such a case is given by cancer cells, which are always capable to maintain telomeres, either expressing telomerase (in more than 90% of the cases) (Counter et al., 1995; Harley, 2008) or using alternative telomere lengthening mechanisms (Muntoni & Reddel, 2005).

Telomerase and telomeric proteins act primarily at telomeres; however, more and more evidence indicates that their role is not restricted at chromosome ends, but they play important functions in different cellular pathways. In this review, we will discuss some of the extra-curricular activities of telomerase and its subunits hTERT and hTR, and of the telomeric protein TRF2.

Ectopic Tert Expression, Cellular Immortalization and Transformation

Ectopic expression of hTERT in different types of somatic cells restores telomerase activity and leads to cellular immortalization (Bodnar et al., 1998; Harley, 2002). Telomere stabilization is a key event in immortalization; however, studies on hTERT immortalized cell lines revealed that telomerase can play additional functions besides telomere lengthening, which can in turn play a role in cellular transformation (reviewed in Belgiovine et al., 2008). One of the important consequences of hTERT ectopic expression is a change in gene expression profiles. In particular, variations in the expression of genes implicated in the regulation of cellular growth have been reported, which can lead to an increased growth rate in hTERT expressing cells. In human mammary epithelial cells (HMEC), telomerase activation has been found to be associated with the induction of a cellular mitogenic program, which correlates with the overexpression of several genes stimulating proliferation (Smith et al.,

2003). One of such genes is the epidermal growth factor receptor (*EGFR)* gene, which is known to be involved in mammary epithelial cell proliferation as well as in breast cancer. This effect on gene expression requires a functioning hTERT; in fact, transduction with a construct coding for a catalytically inactive protein did not cause changes in cellular growth (Smith et al., 2003). Upregulation of *epiregulin*, a gene coding for a potent growth factor belonging to the EGF family, was detected in hTERT immortalized fibroblasts, again leading to increased cellular proliferation (Lindvall et al., 2003).

Resistance to transforming growth factor β (TGF-β) growth inhibition was found in HMEC ectopically expressing hTERT (Stampfer et al., 2001), similarly to what described in carcinogen-treated HMEC that overcame senescence, acquired telomerase activity and became immortal (Garbe et al., 1999). A link between increased growth rate and modulation of the TGF-β signalling network was also found in mouse embryonic fibroblasts (MEFs) expressing exogenous mouse TERT (mTERT) (Geserick et al., 2006). In these cells, mTERT expression results in growth advantage, increased spontaneous immortalization and colony formation in response to oncogenic stress. Despite this growth-promoting effect of mTERT is telomere-length independent, given the very long telomeres of MEFs, it requires catalytically active mTERT, as well as the formation of mTERT/TR complexes. Gene expression profiles of mTERT transduced MEFs showed a reduction in the expression of several genes that are downstream targets of TGF-β, together with other genes that are known to be downregulated in murine and human tumors.

Complex changes in gene expression were also described in hTERT-immortalized human urothelial cells, in which genes involved in stem cell renewal, differentiation and tumorigenesis were differentially expressed as soon as cells were transduced with hTERT (Chapman et al., 2008). In this cellular system, gene expression modulation did not require a catalytically active hTERT. Similarly, hTERT, independently of telomerase activity, is capable of activating VEGF expression in embryonic lung cells and in HeLa cells (Zhou et al., 2009). VEGF expression is modulated at the transcriptional level; in fact, in HeLa cells cotransfected with a luciferase reporter construct containing the VEGF promoter and an expression vector for hTERT, the VEGF promoter shows a four fold greater activity than in HeLa cells cotransfected a vector devoid of hTERT.

Telomerase and Neoplasia

A clear role of hTERT in cellular transformation independently of telomere lengthening was demonstrated by Stewart et al. (2002). These authors showed that cells able to maintain telomeres through a telomerase independent mechanism were immortal, but became able to induce tumours in immunocompromised mice only after ectopic hTERT expression. Even an HA-tagged hTERT, which does not elongate telomeres *in vivo*, was able to induce tumorigenic conversion, definitely confirming that this role in tumorigenesis is independent of telomere elongation.

Mouse models also provided evidence for non-telomeric effects of telomerase in tumour development. Overexpression of mTERT in FVB/N mice enormously augmented tumorigenesis over the lifetime of the animals (Artandi et al., 2002). The transgene was expressed in a broad range of tissues; however, in most cases transgenic mice developed breast cancer, despite the resistance of the FBV strain to this type of tumour, suggesting that the telomerase-induced cancer-prone phenotype is tissue specific. Taking into account that mouse cells have ample telomere reserves, these results are in agreement with the hypothesis that telomerase plays functions beyond its ability to sustain telomere length.

In the mouse model described above, telomerase influences spontaneous tumour susceptibility; however, increased tumour induction upon exposure to chemical carcinogens was also described in mice ectopically expressing the catalytic subunit of telomerase in basal keratinocytes (K5-mTERT mice) (Gonzalez-Suarez et al., 2001). In these mice, ectopic mTERT expression did not result in an extension of telomeres, but was associated with an increased susceptibility in developing papillomas after exposure to chemical carcinogens. Because an increased proliferation of basal keratinocytes was observed after treatment of K5-mTERT mice with chemical carcinogens, as well as a faster rate of re-epithelisation of the skin in wound-healing experiments, it was proposed that ectopic telomerase expression conferred a proliferative advantage, which could be involved in promoting tumour formation and progression. Both mTERT and mTR are required for the increased tumorigenesis; in fact, TR knock down in K5-mTERT relieves mice of tumour induction (Cayuela et al., 2005). Surprisingly, TERT overexpression in the absence of TR had a protective role in relation to tumour formation.

Inhibition of telomerase activity in cancer cells causes telomere shortening, which can lead to cell cycle arrest or death (Harley, 2008). However, a direct action against TERT or TR expression can rapidly drive cells to death without telomere shortening. A rapid growth inhibitory and apoptotic response was observed in cancer cells in which hTERT expression had been knocked down by antisense expression (Folini et al., 2003), indicating that the reduction of the telomerase ribonuclear complex levels has a dramatic effect on cancer cell viability independently of telomerase telomeric effects. Expression of a short-hairpin interfering RNA against hTR, inhibited cell growth and induced apoptosis very rapidly, without telomere shortening (Li et al., 2004). Li et al. (2005) showed that the quick response to TR depletion did not involve bulk telomere shortening or telomere uncapping, nor it induced a DNA damage response or required p53, but it was related to a modulation of gene expression, with a decreased transcription of genes involved in cell cycle progression, tumour growth, angiogenesis and metastasis. Since no effect were observed in telomerase negative cell lines, the authors conclude that downregulation of the telomerase ribonuclear complex in cancer cells is the key event in eliciting the rapid response.

Thus, as well as the induction of telomerase expression in somatic cells can induce change in gene expression and promote cell growth (see previous paragraph), the inhibition of the expression of telomerase components in cancer cells can rapidly elicit growth arrest. In both cases, telomere lengthening or telomere shortening are not required.

Uncoupling a Pair: hTR and hTERT Play on Their Own

hTR Extra-Telomeric Activities

Despite the mechanisms by which telomerase, hTERT and hTR can play a role in cellular transformation independently of telomere length maintenance are still to be defined, new specific functions for hTR and hTERT in different cellular processes are now starting to be highlighted.

Recently, a new function for hTR was discovered in the telomere context, but completely independent of hTERT and telomerase activity. Ting et al. (2009) demonstrated that hTR can stimulate the DNA-dependent protein kinase (DNA-PK) to phosphorylate heterogeneous nuclear ribonucleoprotein A1 (hnRNP A1). HnRNP A1 belongs to a large family of nucleic acid binding proteins implicated in many aspects of RNA/DNA metabolism (for a review see Carpenter et al., 2006). By the way of its two RNA recognition motifs

(RRM 1 and 2), hnRNP A1 directly binds both TR (Fiset & Chabot, 2001) and single or double-stranded telomeric DNA (Dallaire et al., 2000; Zhang et al., 2006). Interestingly, a role for hnRNP A1 in telomere biogenesis was demonstrated by La Branche et al. in 1998; these authors described a mouse cell line deficient in hnRNP A1 that showed telomeres shorter than A1 positive cells; restoring A1 expression in deficient cells was sufficient to increase telomere length (LaBranche et al., 1998). On the basis of this observation, it was then speculated that hnRNP A1 could induce a remodelling of telomeric repeat structure, in order to facilitate telomerase access to telomeres (Ford et al., 2002).

DNA-PK is composed of a catalytic subunit (DNA-PKcs) and a DNA binding domain (the Ku 70/80 heterodimer). This enzyme plays a key role in the non-homologous end-joining (NHEJ) pathway for the repair of DNA double strand breaks (DSBs) (Lees-Miller & Meek, 2003). In addition, the different enzyme subunits, and the kinase activity itself, are involved in telomere metabolism, acting both in the telomere end capping function and in telomere length maintenance (Hsu et al., 2000; Gilley et al., 2001; Bailey et al., 2004; Williams et al., 2009).

Given that hTR can directly interact with the Ku heterodimer (Ting et al., 2005), Ting et al. (2009) wondered whether this interaction could stimulate the kinase activity of DNA-PK. Using an *in vitro* assay, these authors demonstrated that hTR stimulates DNA-PK activity and hnRNP A1 phosphorylation. DNA-PK activation by hTR is highly specific: in fact, while double strand (ds) DNA, the major DNA-PK activator, could trigger kinase activation and phosphorylation of all well-known substrates of DNA-PK, and also of hnRNP A1, the presence of hTR could drive phosphorylation only of hnRNP A1. The reciprocal immunoprecipitation of Ku and hnRNP A1 both from cell lines that express, or do not express, hTR suggests an RNA-independent interaction between these proteins. Nevertheless, the binding of DNA-PK and hnRNP A1 to the same hTR molecule, with formation of a nucleoprotein complex, is necessary to support an efficient phosphorylation of hnRNP A1. Moreover, immunoprecipitation of hnRNP A1 from cells treated with specific DNA-PK inhibitors, or from cells that do not express hTR, revealed a reduced level of phosphorylation of the protein, suggesting that *in vivo* hnRNP A1 is a physiological substrate for DNA-PK.

Which could be the functional role of hTR activation of DNA-PK? Ting et al. (2009) speculates that the interaction of Ku and hnRNP A1 with hTR could recruit and activate DNA-PKcs at telomeres to phosphorylate hnRNP A1 itself, and probably other telomeric proteins, in order to modulate their function during telomere synthesis. One possibility is that phosphorylation of hnRNP A1 could influence its ability to recruit telomerase to chromosome ends or modulate higher order telomere structures.

It is worthwhile pointing out that, whereas both hTERT and telomerase activity are mostly absent in somatic cells, hTR is ubiquitously expressed (Feng et al., 1995), making it very likely that hTR plays other telomerase-independent roles in cell biology. In agreement with this hypothesis, it was recently demonstrated that hTR regulates DNA damage pathways in response to UV irradiation by a feedback mechanism bringing down ATR (ataxia telangectasia related protein) activity to reinitiate cell cycle progression (Kedde et al., 2006). Notably, ATR is a phosphoinositol kinase-like kinase family member related to DNA-PKcs and both are implicated in the same damage response pathway (Yajima et al., 2006). These telomerase-independent functions of hTR provide further insights into the complexity of the relationships among different cellular pathways, given that the same molecule can play independent roles in different biological processes.

hTERT Works as a Transcripton Factor

In 2005, two laboratories independently discovered a novel function for the catalytic subunit of telomerase, i.e. the capability to promote proliferation of quiescent stem cells, in particular, of the multipotent stem cells of hair follicles (Flores et al., 2005; Sarin et al., 2005). Hair growth is regulated by a switching from the resting phase of the hair follicle cycle (telogen) to the active phase (anagen), determined by activation of a limited number of stem cells that reside in the bulge region, a niche at the follicle base (Tumbar et al., 2004). Sarin et al. (2005) demonstrated that TERT expression in adult skin is sufficient to initiate a transition in the hair follicles from the telogen to the anagen phase, inducing proliferation of quiescent stem cells. Importantly, this function does not require TR and telomere synthesis (Sarin et al., 2005). However, critically short telomeres inhibit the mobilization of epidermal stem cells, reducing the proliferative potential of the hair follicle (Flores et al., 2005; Sarin et al., 2005; Siegl-Cachedenier et al., 2007). These observations point to a double role for TERT in stem cell biology, both through telomere length maintenance and through the regulation of stem cell proliferation. Recently, comparative analysis of genome-wide transcriptional responses revealed that both TERT and its catalytically inactive mutant (TERT^{ci}) trigger a rapid change in gene expression that significantly overlaps the transcriptional program regulated by Wnt (Choi et al., 2008). Thus, it is likely that TERT activates epidermal progenitor not through its reverse transcriptase activity, but as a key factor in the Wnt cascade.

It is known that Wnt proteins can act as stem cell growth factors, promoting the maintenance and proliferation of stem cells, among which, epidermal stem cells (Van Mater et al., 2003; Reya & Clevers, 2005). Wnt proteins prime the so-called Wnt cascade, binding to cell surface receptors and leading to the accumulation of β-catenin in the nucleus. In the nucleus, β-catenin interacts with DNA binding proteins of the Tcf/Lef family (Eastman & Grosschedl, 1999), promoting their transcriptional activator function (Reya & Clevers, 2005).

An extraordinary improvement for the comprehension of cellular signalling pathways was represented by the finding that TERT directly modulates the Wnt pathway by acting as a transcription factor in β-catenin complexes (Park et al., 2009). Park et al. (2009) found that TERT interacts with BRG1, a chromatin remodeller that interacts with β-catenin and is involved in the Wnt signalling (Barker et al., 2001). Subsequently, they discovered that TERT overexpression was able to activate reporter plasmids driven by promoters of several β-catenin target genes and this activation was strictly dependent on BRG1 expression. Finally, by sequential chromatin immunoprecipitations, they demonstrated that TERT not only was bound to promoters responsive to Wnt signalling, but also that the binding happened exactly on the same promoter elements recognized by BRG1 and β-catenin.

To test whether TERT could carry out transcription activation *in vivo*, Park et al. (2009) created a transgenic mouse crossing tetracycline-inducible TERT^{ci} mice with a knock-in strain in which β-galactosidase (β-gal) was expressed from the Axin2 promoter, a well-known target of the Wnt signalling (Lustig et al., 2002). Monitoring β-gal activity in crypts of the small intestine and colon of these mice revealed that this was enhanced upon induction of catalytically inactive TERT. On the other hand, deletion of TERT in conditional knockout ES cells significantly diminished induction of endogenous Axin2 and reduced its basal expression. These findings demonstrate that TERT is both necessary and sufficient to sustain the transcriptional output of the Wnt pathway. Abnormal activation of Wnt pathway in *Xenopus* embryos causes duplication of the anterior-posterior axis (McMahon & Moon, 1989). Injecting increasing amounts of *Xenopus* TERT (xTERT) mRNA promotes formation of a duplicate anterior-posterior axis in a dose-dependent manner, whereas xTERT depletion causes aberrations in axis formation. Moreover, xTERT inhibition provokes ectopic neural tube formation and loss of the notochord, phenotypes also seen in knockout mouse embryos defective in components of the Wnt pathway (Yoshikawa et al., 1997); in addition, xTERT depletion blocks the induction of the expression of *Cdx* genes, which belong to Wnt responsive gene group (Ikeya & Takada, 2001). Thus, TERT expression is required both for proper development and for homeotic gene expression regulation. Curiously, G1 Tert$^{-/-}$ mice, which lacks telomerase activity, were considered phenotipically normal (Rajaraman et al., 2007), but a high-resolution CT-scanning revealed a homeotic transformation of the vertebrae (Park et al., 2009).

These data reveal a novel function for the catalytic subunit of telomerase distinguished from telomerase activity, in fact, TERT works also as an essential transcription factor engaged in the last step of Wnt signalling pathway.

hTERT Betrays hTR: It Meets Another Partner

Purification of human telomerase complexes revealed an unexpected heterogeneity in protein composition (Fu & Collins, 2007). Recently, Maida et al. (2009) showed that hTERT can also interact with different RNA molecules; in particular, they found that a heterogeneous mixture of 38 RNA sequences can associate with hTERT, 5% of which corresponded to hTR and to the RNA component of mitochondrial RNA processing endoribonuclease (RMRP). TERT-TR and TERT-RMRP complexes showed similar abundance, although TR is expressed at five-fold higher levels than RMRP. RMRP is a 267-nucleotide noncoding RNA, it is encoded by a nuclear single-copy gene and is then imported into mitochondria (Hsieh et al., 1990). Mutations in the *RMRP* gene are responsible for cartilage-hair hypoplasia (CHH), an autosomal recessive disorder mainly characterized by short stature, hypoplastic hair, defective immunity and hypoplastic anaemia (Ridanpaa et al., 2001). All *RMRP* gene mutations lead to decreased cell growth by impairing ribosomal assembly and by altering cyclin-dependent cell cycle regulation (Thiel et al., 2005).

An *in vitro* assay performed incubating TERT with either one or the other recombinantRNA ruled out the possibility that RMRP could substitute TR to reconstitute telomerase activity. On the other hand, the TERT-RMRP complex, but not the TERT-TR one, clearly had an RNA-dependent RNA polymerase (RdRP) activity, as demonstrated by an RNA synthesis assay. The product of this reaction is an RNA molecule composed of sense and antisense strands of RMRP that folds in a double-stranded hairpin structure. Interestingly, a catalytically inactive TERT mutant, that fails to elongate telomeres (Masutomi et al., 2003), retains the ability to bind RMRP, but the complex lacks detectable RdRP activity. Thus TERT acts as the catalytic subunit for both the telomerase reverse transcriptase and the RNA-dependent RNA polymerase.

Overexpression of RMRP in cell lines that lack or constitutively express TERT led to uncover the relationship between RMRP levels and RdRP activity. In particular, overexpression of RMRP in cells lacking TERT resulted in accumulation of RMRP itself; whereas in cells that express TERT, the RMRP steady-state levels were decreased. These findings suggest that RMRP levels are dependent on RdRP activity by a feedback mechanism. By northern blotting analysis, it was found that both sense and antisense probes corresponding to RMRP hybridised to double-stranded 22-nucleotide RNAs, that are *bona fide* short interfering RNAs (siRNAs). Suppression of Dicer, that catalyzes the first step in the RNA interference pathway and initiates formation of the RNA-induced silencing complex (RISC) (Jaronczyk et al., 2005), led to diminished levels of these 22-nt RNAs, confirming that dsRNA molecules produced by RdRP activity of the TERT-RMRP complex are processed into siRNAs. Moreover, only the sense strands of the endogenous RMRP-specific siRNA were associated with human AGO2, demonstrating that they are processed by RISC. Thus, as described for other cellular RdRPs, also the TERT-RMRP RdRP synthesizes dsRNA precursors for siRNA, that regulate the expression of RMRP *in vivo*. The demonstration that TERT and RMRP form a peculiar ribonucleoprotein complex able to produce dsRNAs further processed in siRNAs implies again an involvement of TERT in gene expression regulation, although with a mechanism completely different from that discussed in the previuos paragraph. Other RNAs different from RMRP that can act as templates for the TERT-RMRP RdRP remain to be identified; however, it cannot be excluded that the TERT-RMRP complex regulates the expression of other genes by generating siRNA.

It is well known that hTERT is implicated in dyskeratosis congenita, a syndrome characterized by stem cell failure (Kirwan & Dokal, 2009), because of its role in telomere maintenance; since mutations in RMRP are found in CHH, an appealing possibility is that perturbations of the TERT-RMRP complex may be a crucial factor in the the pathogenesis of this disorder. These findings suggest that TERT can form different complexes that play critical roles in multiple aspects of stem cell biology.

The data reported above indicate that hTERT, besides being a reverse transcriptase in association with hTR, can have an RNA-dependent RNA polymerase activity, in association with RMRP. Lue et al. (2005) discovered another enzymatic activity of TERT; they found that, in the presence of particular metal ion concentrations, TERT can switch to a template- and RNA-independent mode of DNA synthesis. This hitherto terminal transferase activity depends on the active site of the telomerase catalytic subunit but it does not require TR.

TRF2 beyond Telomeres: Its Involvement in DNA Damage Response and Repair

As well as for hTERT and hTR, also for TRF2 several functions besides its role at telomeres have been ascertained. TRF2 was discovered as a telomeric protein on the basis of its sequence similarity to TRF1, the first TTAGGG binding protein identified (Bilaud et al., 1997; Broccoli et al., 1997). Three major telomeric functions are attributed to TRF2: 1) it plays a role in controlling telomere length; 2) it represses the ATM kinase pathway at telomeres; 3) it prevents NHEJ between telomeres. TRF2 probably performs the last two tasks by facilitating telomere organization into t-loops, which hide telomeric DNA ends from repair enzymes (Figure 1A) (de Lange, 2009). Loss of TRF2 at telomeres causes the loss of the 3' telomeric overhang, telomere end fusions and an ATM- and p53 mediated DNA damage response, which triggers apoptosis or senescence (van Steensel et al., 1998). The function of TRF2 at telomeres is clearly fundamental for the maintenance of genome stability; cumulating evidence indicates that TRF2 plays also a general role in the response to DNA damage in genomic sites outside of telomeres, being both involved in resistance to DNA damaging agents and participating to repair processes.

An early evidence showing that TRF2 is involved in the response to DNA damage was the observation that TRF2 levels increase upon treatments with DSB inducers, such as the topoisomerase inhibitor etoposide, or irradiation (Klapper et al., 2003; Bradshaw et al., 2005). Bradshow et al. (2005) created DSBs in defined regions of the nucleus by exposing the cells to the DNA intercalating agent Hoechst 33258 and pulsed UVA laser microbeam irradiation; they found that within 2 seconds of irradiation, TRF2 associated with DNA breaks forming transient foci. Foci formation did not require ATM or other DNA repair proteins such as DNA-PKcs, MRE11/Rad50/NBS1 complex, Ku70, Werner and Bloom proteins. The irradiation scheme used by Bradshow and collegues induces a high local concentration of DSBs, which is probably required to permit the visualization of TRF2 foci at the site of DNA damage. In fact, cell exposure to lower intensity sources of radiations is not associated with disctinct TRF2 foci formation (Williams et al., 2007). However, it cannot also be excluded that accumulation of TRF2 at damage sites depends on the specific source or type of damage.

Increased TRF2 expression was found at early time after exposure of multidrug resistant (MDR) gastric cancer cells to DSB inducers, such as etoposide or adriamycin (Ning et al., 2006). In MDR cell lines, TRF2 up-regulation occurred at higher levels and earlier than in parental cells and preceded the induction of DNA damage response proteins like ATM, γH2AX and p53, which was greater in parental cells than in MDR cells. These observations suggest that TRF2 might promote drug resistance in cancer cells by interfering with the ATM-mediated DNA damage response induced by damaging agents. In support to this hypothesis, the authors showed that downregulation of TRF2 expression increased drug sensitivity in MDR cells, while TRF2 overexpression in gastric cancer cells increased resistance to different anticancer drugs, mimicking the multidrug resistance phenotype.

Overexpression of TRF2 protects also by DSBs and telomere erosion induced by the DNA damaging agent salvicidine (Zhang et al., 2008). Exposure of cancer cells to salvicidine does not affect TRF2 expression at the RNA levels, but lowers the levels of the protein, probably determining its degradation. Since salvicidine treatment significantly decreases TRF2 bound to telomeres, it was proposed the DNA damaging action of this drug, both at telomeres and at different genomic sites, is potentiated by the destruction of the genome protection mechanisms based on TRF2. Indeed, TRF2 downregulation by RNA interference resulted in a higher level of DSB induction and in a more severe erosion of telomeres. Taken together, these results strongly indicate that TRF2 plays a role in the protection of genomic DNA from genotoxic insults.

TRF2 has been shown to interact, both *in vitro* and *in vivo*, with different proteins involved in base excision repair (BER) of oxidative damage, such as the Werner protein, the FEN1 endonuclease and DNA polymerase β (Fotiadou et al., 2004; Muftuoglu et al., 2006). In particular, TRF2 stimulates DNA polymerase β *in vitro* activity both on telomeric and non-telomeric substrates, suggesting that TRF2 may enhance the repair of damaged bases, in particular the long patch BER, which requires the synthesis of longer DNA tracts.

A clear evidence of TRF2 involvement in DNA repair is its rapid phosphorylation in response to DSBs (Tanaka et al., 2005). TRF2 phosphorylation depends on ATM. The phosphorylated form of the protein is transient; in fact, it is detected rapidly at DNA damage sites after irradiation, but it is also dissipated very rapidly, suggesting that it functions at early stages of DNA damage response. Phosphorylated TRF2 is not normally present at telomeres, while it can be detected at telomeres in cells undergoing telomere-based crisis, indicating that it associates only with dysfunctional telomeres that have induced a localized DNA damage response.

Recently, Huda et al. (2009) have shown that DNA damage-induced phosphorylation of TRF2 plays a functional role in the repair of DNA DSBs. These authors found that overexpression of a TRF2 mutant protein that cannot be phosphorylated in tumor cells causes X-ray hypersensitivity, which is not due to altered apoptosis or telomere shortening, but to a defect in the repair of DSBs. Repair of DSBs after X-ray exposure follows a bimodal kinetics, with a fast and a low repair pathway (Iliakis et al., 2004). By comet assay analysis, Huda et al. (2009) demonstrated that, in the absence of TRF2 phosphorylation, the fast removal of DSBs did not occur, while the slow DSB repair pathway was unaffected. In agreement with these results, the analysis of the kinetics of γ-H2AX foci formation after X-ray exposure showed that, in the absence of TRF2 phosphorylation, the highest peak levels are reached approximately 1 hour after irradiation (similarly to wild type cells), while the reduction of γ-H2AX to steady state levels is greatly delayed compared to wild type cells. How the phosphorylated form of TRF2 participate to DBS repair is a matter of investigation. One possibility envisaged by Huda et al. (2009) is that phosphorylation of TRF2 either facilitates or disrupts specific protein-protein interactions at DSB sites; alternatively, or in addition, TRF2 phosphorylation could facilitate TRF2 dissociation from telomeres and its localization at DNA damage sites.

The results reported above suggest that TRF2 phosphorylation is involved in the repair of DSBs via the non homologous end-joining pathway. Using a different experimental system, based on reporter cassettes integrated into a chromosomal locus and allowing the detection of either NHEJ or homologous recombination (HR) events after DNA damage induction, Mao et al. (2007) found that TRF2 overexpression in hTERT immortalized primary fibroblasts inhibits NHEJ events, while stimulates HR.

Conclusion

It is now clear that the main telomerase components, TERT and TR, as well as an important telomeric protein such as TRF2, play specific roles in different cellular processes that are not related to telomere metabolism. Particularly interesting is the discovery of the ability of TERT to work as a transcription factor in the Wnt signalling pathway (Park et al. 2009); it might be speculated that TERT could play a direct role in the transcription of other genes, which are known to be modulated upon induction of TERT expression.

A considerable number of reports show that TRF2 is involved in DNA damage response and DNA repair upon damage induction in genomic sites different from telomeres, indicating that, as well as DNA repair proteins play a role at telomeres, telomeric proteins can be utilized for repair pathways. To this regard, it is also worth mentioning the possible involvement of telomerase in DSBs repair described for the genesis of intertitial telomeric repeats during evolution (Nergadze et al. 2007).

Telomerase and telomeric proteins are considered potential targets for cancer therapy; the understanding of the molecular mechanisms underlying their extra-telomeric functions is fundamental to be able to design the best targeting strategy and to unravel possible side effects.

Acknowledgments

We are grateful to A. Ivana Scovassi for critical reading of the manuscript. Work in C.M. laboratory is supported by Fondazione Cariplo (Grant 2006-0734). C.B. is a PhD student of the University of Pavia (Dottorato in Scienze Genetiche e Biomolecolari). I.C. post-doctoral fellowship is supported by Fondazione Cariplo.

References

Artandi, SE; Alson, S; Tietze, MK; Sharpless, NE; Ye, S; Greenberg, RA; Castrillon, DH; Horner, JW; Weiler, SR; Carrasco, RD; DePinho, RA. Constitutive telomerase expression promotes mammary carcinomas in aging mice. *Proc Natl Acad Sci U S A*, 2002, 99, 8191-8196.

Bailey, SM; Brenneman, MA; Halbrook, J; Nickoloff, JA; Ullrich, RL; Goodwin, EH. The kinase activity of DNA-PK is required to protect mammalian telomeres. *DNA Repair (Amst)*, 2004, 3, 225-233.

Barker, N; Hurlstone, A; Musisi, H; Miles, A; Bienz, M; Clevers, H. The chromatin remodelling factor Brg-1 interacts with beta-catenin to promote target gene activation. *Embo J*, 2001, 20, 4935-4943.

Belgiovine, C; Chiodi, I; Mondello, C. Telomerase: cellular immortalization and neoplastic transformation. Multiple functions of a multifaceted complex. *Cytogenet Genome Res.*, 2008, 122, 255-262.

Bilaud, T; Brun, C; Ancelin, K; Koering, CE; Laroche, T; Gilson, E. Telomeric localization of TRF2, a novel human telobox protein. *Nat Genet*, 1997, 17, 236-239.

Blackburn, EH. Telomeres and telomerase: their mechanisms of action and the effects of altering their functions. *FEBS Lett.*, 2005, 579, 859-862.

Bodnar, AG; Ouellette, M; Frolkis, M; Holt, SE; Chiu, CP; Morin, GB; Harley, CB; Shay, JW; Lichtsteiner, S; Wright, WE. Extension of life-span by introduction of telomerase into normal human cells. *Science*, 1998, 279, 349-352.

Bradshaw, PS; Stavropoulos, DJ; Meyn, MS. Human telomeric protein TRF2 associates with genomic double-strand breaks as an early response to DNA damage. *Nat Genet*, 2005, 37, 193-197.

Broccoli, D; Smogorzewska, A; Chong, L; de Lange, T. Human telomeres contain two distinct Myb-related proteins, TRF1 and TRF2. *Nat Genet*, 1997, 17, 231-235.

Campisi, J; d'Adda di Fagagna, F. Cellular senescence: when bad things happen to good cells. *Nat Rev Mol Cell Biol*, 2007, 8, 729-740.

Carpenter, B; MacKay, C; Alnabulsi, A; MacKay, M; Telfer, C; Melvin, WT; Murray, GI. The roles of heterogeneous nuclear ribonucleoproteins in tumour development and progression. *Biochim Biophys Acta*, 2006, 1765, 85-100.

Cayuela, ML; Flores, JM; Blasco, MA. The telomerase RNA component Terc is required for the tumour-promoting effects of Tert overexpression. *EMBO Rep.*, 2005, 6, 268-274.

Chapman, EJ; Kelly, G; Knowles, MA. Genes involved in differentiation, stem cell renewal, and tumorigenesis are modulated in telomerase-immortalized human urothelial cells. *Mol Cancer Res.*, 2008, 6, 1154-1168.

Choi, J; Southworth, LK; Sarin, KY; Venteicher, AS; Ma, W; Chang, W; Cheung, P; Jun, S; Artandi, MK; Shah, N; Kim, SK; Artandi, SE. TERT promotes epithelial proliferation through transcriptional control of a Myc- and Wnt-related developmental program. *PLoS Genet*, 2008, 4, e10.

Cohen, SB; Graham, ME; Lovrecz, GO; Bache, N; Robinson, PJ; Reddel, RR. Protein composition of catalytically active human telomerase from immortal cells. *Science*, 2007, 315, 1850-1853.

Collins, K. Physiological assembly and activity of human telomerase complexes. *Mech Ageing Dev*, 2008, 129, 91-98.

Counter, CM; Gupta, J; Harley, CB; Leber, B; Bacchetti, S. Telomerase activity in normal leukocytes and in hematologic malignancies. *Blood*, 1995, 85, 2315-2320.

Dallaire, F; Dupuis, S; Fiset, S; Chabot, B. Heterogeneous nuclear ribonucleoprotein A1 and UP1 protect mammalian telomeric repeats and modulate telomere replication in vitro. *J Biol Chem.*, 2000, 275, 14509-14516.

de Lange, T. Shelterin: the protein complex that shapes and safeguards human telomeres. *Genes Dev.*, 2005, 19, 2100-2110.

de Lange, T. How telomeres solve the end-protection problem. *Science*, 2009, 326, 948-952.

Eastman, Q; Grosschedl, R. Regulation of LEF-1/TCF transcription factors by Wnt and other signals. *Curr Opin Cell Biol.*, 1999, 11, 233-240.

Feng, J; Funk, WD; Wang, SS; Weinrich, SL; Avilion, AA; Chiu, CP; Adams, RR; Chang, E; Allsopp, RC; Yu, J; et al. The RNA component of human telomerase. *Science*, 1995, 269, 1236-1241.

Fiset, S; Chabot, B. hnRNP A1 may interact simultaneously with telomeric DNA and the human telomerase RNA in vitro. *Nucleic Acids Res.*, 2001, 29, 2268-2275.

Flores, I; Cayuela, ML; Blasco, MA. Effects of telomerase and telomere length on epidermal stem cell behavior. *Science*, 2005, 309, 1253-1256.

Folini, M; Berg, K; Millo, E; Villa, R; Prasmickaite, L; Daidone, MG; Benatti, U; Zaffaroni, N. Photochemical internalization of a peptide nucleic acid targeting the catalytic subunit of human telomerase. *Cancer Res.*, 2003, 63, 3490-3494.

Ford, LP; Wright, WE; Shay, JW. A model for heterogeneous nuclear ribonucleoproteins in telomere and telomerase regulation. *Oncogene*, 2002, 21, 580-583.

Fotiadou, P; Henegariu, O; Sweasy, JB. DNA polymerase beta interacts with TRF2 and induces telomere dysfunction in a murine mammary cell line. *Cancer Res.*, 2004, 64, 3830-3837.

Fu, D; Collins, K. Purification of human telomerase complexes identifies factors involved in telomerase biogenesis and telomere length regulation. *Mol Cell*, 2007, 28, 773-785.

Garbe, J; Wong, M; Wigington, D; Yaswen, P; Stampfer, MR. Viral oncogenes accelerate conversion to immortality of cultured conditionally immortal human mammary epithelial cells. *Oncogene*, 1999, 18, 2169-2180.

Geserick, C; Tejera, A; Gonzalez-Suarez, E; Klatt, P; Blasco, MA. Expression of mTert in primary murine cells links the growth-promoting effects of telomerase to transforming growth factor-beta signaling. *Oncogene*, 2006, 25, 4310-4319.

Gilley, D; Tanaka, H; Hande, MP; Kurimasa, A; Li, GC; Oshimura, M; Chen, DJ. DNA-PKcs is critical for telomere capping. *Proc Natl Acad Sci U S A*, 2001, 98, 15084-15088.

Gonzalez-Suarez, E; Samper, E; Ramirez, A; Flores, JM; Martin-Caballero, J; Jorcano, JL; Blasco, MA. Increased epidermal tumors and increased skin wound healing in transgenic mice overexpressing the catalytic subunit of telomerase, mTERT, in basal keratinocytes. *Embo J*, 2001, 20, 2619-2630.

Greider, CW; Blackburn, EH. Identification of a specific telomere terminal transferase activity in Tetrahymena extracts. *Cell*, 1985, 43, 405-413.

Griffith, JD; Comeau, L; Rosenfield, S; Stansel, RM; Bianchi, A; Moss, H; de Lange, T. Mammalian telomeres end in a large duplex loop. *Cell*, 1999, 97, 503-514.

Harley, CB. Telomere loss: mitotic clock or genetic time bomb? *Mutat Res.*, 1991, 256, 271-282.

Harley, CB. Telomerase is not an oncogene. *Oncogene*, 2002, 21, 494-502.

Harley, CB. Telomerase and cancer therapeutics. *Nat Rev Cancer*, 2008, 8, 167-179.

Harley, CB; Futcher, AB; Greider, CW. Telomeres shorten during ageing of human fibroblasts. *Nature*, 1990, 345, 458-460.

Hsieh, CL; Donlon, TA; Darras, BT; Chang, DD; Topper, JN; Clayton, DA; Francke, U. The gene for the RNA component of the mitochondrial RNA-processing endoribonuclease is located on human chromosome 9p and on mouse chromosome 4. *Genomics*, 1990, 6, 540-544.

Hsu, HL; Gilley, D; Galande, SA; Hande, MP; Allen, B; Kim, SH; Li, GC; Campisi, J; Kohwi-Shigematsu, T; Chen, DJ. Ku acts in a unique way at the mammalian telomere to prevent end joining. *Genes Dev*, 2000, 14, 2807-2812.

Huda, N; Tanaka, H; Mendonca, MS; Gilley, D. DNA damage-induced phosphorylation of TRF2 is required for the fast pathway of DNA double-strand break repair. *Mol Cell Biol.*, 2009, 29, 3597-3604.

Ikeya, M; Takada, S. Wnt-3a is required for somite specification along the anteroposterior axis of the mouse embryo and for regulation of cdx-1 expression. *Mech Dev*, 2001, 103, 27-33.

Iliakis, G; Wang, H; Perrault, AR; Boecker, W; Rosidi, B; Windhofer, F; Wu, W; Guan, J; Terzoudi, G; Pantelias, G. Mechanisms of DNA double strand break repair and chromosome aberration formation. *Cytogenet Genome Res.*, 2004, 104, 14-20.

Jaronczyk, K; Carmichael, JB; Hobman, TC. Exploring the functions of RNA interference pathway proteins: some functions are more RISCy than others? *Biochem J*, 2005, 387, 561-571.

Kedde, M; le Sage, C; Duursma, A; Zlotorynski, E; van Leeuwen, B; Nijkamp, W; Beijersbergen, R; Agami, R. Telomerase-independent regulation of ATR by human telomerase RNA. *J Biol Chem*, 2006, 281, 40503-40514.

Kirwan, M; Dokal, I. Dyskeratosis congenita, stem cells and telomeres. *Biochim Biophys Acta*, 2009, 1792, 371-379.

Klapper, W; Qian, W; Schulte, C; Parwaresch, R. DNA damage transiently increases TRF2 mRNA expression and telomerase activity. *Leukemia*, 2003, 17, 2007-2015.

LaBranche, H; Dupuis, S; Ben-David, Y; Bani, MR; Wellinger, RJ; Chabot, B. Telomere elongation by hnRNP A1 and a derivative that interacts with telomeric repeats and telomerase. *Nat Genet*, 1998, 19, 199-202.

Lees-Miller, SP; Meek, K. Repair of DNA double strand breaks by non-homologous end joining. *Biochimie*, 2003, 85, 1161-1173.

Li, S; Crothers, J; Haqq, CM; Blackburn, EH. Cellular and gene expression responses involved in the rapid growth inhibition of human cancer cells by RNA interference-mediated depletion of telomerase RNA. *J Biol Chem*, 2005, 280, 23709-23717.

Li, S; Rosenberg, JE; Donjacour, AA; Botchkina, IL; Hom, YK; Cunha, GR; Blackburn, EH. Rapid inhibition of cancer cell growth induced by lentiviral delivery and expression of mutant-template telomerase RNA and anti-telomerase short-interfering RNA. *Cancer Res.*, 2004, 64, 4833-4840.

Lindvall, C; Hou, M; Komurasaki, T; Zheng, C; Henriksson, M; Sedivy, JM; Bjorkholm, M; Teh, BT; Nordenskjold, M; Xu, D. Molecular characterization of human telomerase reverse transcriptase-immortalized human fibroblasts by gene expression profiling: activation of the epiregulin gene. *Cancer Res.*, 2003, 63, 1743-1747.

Lue, NF; Bosoy, D; Moriarty, TJ; Autexier, C; Altman, B; Leng, S. Telomerase can act as a template- and RNA-independent terminal transferase. *Proc Natl Acad Sci U S A*, 2005, 102, 9778-9783.

Lustig, B; Jerchow, B; Sachs, M; Weiler, S; Pietsch, T; Karsten, U; van de Wetering, M; Clevers, H; Schlag, PM; Birchmeier, W; Behrens, J. Negative feedback loop of Wnt signaling through upregulation of conductin/axin2 in colorectal and liver tumors. *Mol Cell Biol.*, 2002, 22, 1184-1193.

Maida, Y; Yasukawa, M; Furuuchi, M; Lassmann, T; Possemato, R; Okamoto, N; Kasim, V; Hayashizaki, Y; Hahn, WC; Masutomi, K. An RNA-dependent RNA polymerase formed by TERT and the RMRP RNA. *Nature*, 2009, 461, 230-235.

Mao, Z; Seluanov, A; Jiang, Y; Gorbunova, V. TRF2 is required for repair of nontelomeric DNA double-strand breaks by homologous recombination. *Proc Natl Acad Sci U S A*, 2007, 104, 13068-13073.

Masutomi, K; Yu, EY; Khurts, S; Ben-Porath, I; Currier, JL; Metz, GB; Brooks, MW; Kaneko, S; Murakami, S; DeCaprio, JA; Weinberg, RA; Stewart, SA; Hahn, WC. Telomerase maintains telomere structure in normal human cells. *Cell*, 2003, 114, 241-253.

McMahon, AP; Moon, RT. int-1--a proto-oncogene involved in cell signalling. *Development*, 1989, 107 Suppl, 161-167.

Muftuoglu, M; Wong, HK; Imam, SZ; Wilson, DM, 3rd; Bohr, VA; Opresko, PL. Telomere repeat binding factor 2 interacts with base excision repair proteins and stimulates DNA synthesis by DNA polymerase beta. *Cancer Res.*, 2006, 66, 113-124.

Muller, H. The remaking of chromosomes. *Collecting Net*, 1938, 13, 181-198.

Muntoni, A; Reddel, RR. The first molecular details of ALT in human tumor cells. *Hum Mol Genet*, 2005, 14 Spec No. 2, R191-196.

Nergadze, SG; Santagostino, MA; Salzano, A; Mondello C; Giulotto E. Contribution of telomerase RNA retrotranscription to DNA double-strand break repair during mammalian genome evolution. *Genome Biol.*, 2007, 8, R260-

Ning, H; Li, T; Zhao, L; Li, T; Li, J; Liu, J; Liu, Z; Fan, D. TRF2 promotes multidrug resistance in gastric cancer cells. *Cancer Biol Ther.*, 2006, 5, 950-956.

Palm, W; de Lange, T. How shelterin protects mammalian telomeres. *Annu Rev Genet*, 2008, 42, 301-334.

Park, JI; Venteicher, AS; Hong, JY; Choi, J; Jun, S; Shkreli, M; Chang, W; Meng, Z; Cheung, P; Ji, H; McLaughlin, M; Veenstra, TD; Nusse, R; McCrea, PD; Artandi, SE. Telomerase modulates Wnt signalling by association with target gene chromatin. *Nature*, 2009, 460, 66-72.

Rajaraman, S; Choi, J; Cheung, P; Beaudry, V; Moore, H; Artandi, SE. Telomere uncapping in progenitor cells with critical telomere shortening is coupled to S-phase progression in vivo. *Proc Natl Acad Sci U S A*, 2007, 104, 17747-17752.

Reya, T; Clevers, H. Wnt signalling in stem cells and cancer. *Nature*, 2005, 434, 843-850.

Ridanpaa, M; van Eenennaam, H; Pelin, K; Chadwick, R; Johnson, C; Yuan, B; vanVenrooij, W; Pruijn, G; Salmela, R; Rockas, S; Makitie, O; Kaitila, I; de la Chapelle, A. Mutations in the RNA component of RNase MRP cause a pleiotropic human disease, cartilage-hair hypoplasia. *Cell*, 2001, 104, 195-203.

Sarin, KY; Cheung, P; Gilison, D; Lee, E; Tennen, RI; Wang, E; Artandi, MK; Oro, AE; Artandi, SE. Conditional telomerase induction causes proliferation of hair follicle stem cells. *Nature*, 2005, 436, 1048-1052.

Shampay, J; Szostak, JW; Blackburn, EH. DNA sequences of telomeres maintained in yeast. *Nature*, 1984, 310, 154-157.

Siegl-Cachedenier, I; Flores, I; Klatt, P; Blasco, MA. Telomerase reverses epidermal hair follicle stem cell defects and loss of long-term survival associated with critically short telomeres. *J Cell Biol.*, 2007, 179, 277-290.

Smith, LL; Coller, HA; Roberts, JM. Telomerase modulates expression of growth-controlling genes and enhances cell proliferation. *Nat. Cell Biol.*, 2003, 5, 474-479.

Stampfer, MR; Garbe, J; Levine, G; Lichtsteiner, S; Vasserot, AP; Yaswen, P. Expression of the telomerase catalytic subunit, hTERT, induces resistance to transforming growth factor beta growth inhibition in p16INK4A(-) human mammary epithelial cells. *Proc Natl Acad Sci USA*, 2001, 98, 4498-4503.

Stewart, SA; Hahn, WC; O'Connor, BF; Banner, EN; Lundberg, AS; Modha, P; Mizuno, H; Brooks, MW; Fleming, M; Zimonjic, DB; Popescu, NC; Weinberg, RA. Telomerase contributes to tumorigenesis by a telomere length-independent mechanism. *Proc. Natl. Acad. Sci. USA*, 2002, 99, 12606-12611.

Szostak, JW; Blackburn, EH. Cloning yeast telomeres on linear plasmid vectors. *Cell*, 1982, 29, 245-255.

Tanaka, H; Mendonca, MS; Bradshaw, PS; Hoelz, DJ; Malkas, LH; Meyn, MS; Gilley, D. DNA damage-induced phosphorylation of the human telomere-associated protein TRF2. *Proc Natl Acad Sci USA*, 2005, 102, 15539-15544.

Thiel, CT; Horn, D; Zabel, B; Ekici, AB; Salinas, K; Gebhart, E; Ruschendorf, F; Sticht, H; Spranger, J; Muller, D; Zweier, C; Schmitt, ME; Reis, A; Rauch, A. Severely incapacitating mutations in patients with extreme short stature identify RNA-processing endoribonuclease RMRP as an essential cell growth regulator. *Am J Hum Genet*, 2005, 77, 795-806.

Ting, NS; Pohorelic, B; Yu, Y; Lees-Miller, SP; Beattie, TL. The human telomerase RNA component, hTR, activates the DNA-dependent protein kinase to phosphorylate heterogeneous nuclear ribonucleoprotein A1. *Nucleic Acids Res.*, 2009, 37, 6105-6115.

Ting, NS; Yu, Y; Pohorelic, B; Lees-Miller, SP; Beattie, TL. Human Ku70/80 interacts directly with hTR, the RNA component of human telomerase. *Nucleic Acids Res*, 2005, 33, 2090-2098.

Tumbar, T; Guasch, G; Greco, V; Blanpain, C; Lowry, WE; Rendl, M; Fuchs, E. Defining the epithelial stem cell niche in skin. *Science*, 2004, 303, 359-363.

Van Mater, D; Kolligs, FT; Dlugosz, AA; Fearon, ER. Transient activation of beta -catenin signaling in cutaneous keratinocytes is sufficient to trigger the active growth phase of the hair cycle in mice. *Genes Dev*, 2003, 17, 1219-1224.

van Steensel, B; Smogorzewska, A; de Lange, T. TRF2 protects human telomeres from end-to-end fusions. *Cell*, 1998, 92, 401-413.

Williams, ES; Klingler, R; Ponnaiya, B; Hardt, T; Schrock, E; Lees-Miller, SP; Meek, K; Ullrich, RL; Bailey, SM. Telomere dysfunction and DNA-PKcs deficiency: characterization and consequence. *Cancer Res.*, 2009, 69, 2100-2107.

Williams, ES; Stap, J; Essers, J; Ponnaiya, B; Luijsterburg, MS; Krawczyk, PM; Ullrich, RL; Aten, JA; Bailey, SM. DNA double-strand breaks are not sufficient to initiate recruitment of TRF2. *Nat Genet*, 2007, 39, 696-698, author reply 698-699.

Yajima, H; Lee, KJ; Chen, BP. ATR-dependent phosphorylation of DNA-dependent protein kinase catalytic subunit in response to UV-induced replication stress. *Mol Cell Biol.*, 2006, 26, 7520-7528.

Yoshikawa, Y; Fujimori, T; McMahon, AP; Takada, S. Evidence that absence of Wnt-3a signaling promotes neuralization instead of paraxial mesoderm development in the mouse. *Dev Biol.*, 1997, 183, 234-242.

Zhang, QS; Manche, L; Xu, RM; Krainer, AR. hnRNP A1 associates with telomere ends and stimulates telomerase activity. *Rna*, 2006, 12, 1116-1128.

Zhang, YW; Zhang, ZX; Miao, ZH; Ding, J. The telomeric protein TRF2 is critical for the protection of A549 cells from both telomere erosion and DNA double-strand breaks driven by salvicine. *Mol Pharmacol*, 2008, 73, 824-832.

Zhou, L; Zheng, D; Wang, M; Cong, YS. Telomerase reverse transcriptase activates the expression of vascular endothelial growth factor independent of telomerase activity. *Biochem Biophys Res Commun*, 2009, 386, 739-743.

In: Telomerase: Composition, Functions ...
Editor: Aiden N. Gagnon, pp. 59-79
ISBN: 978-1-61668-957-5
© 2010 Nova Science Publishers, Inc.

Chapter III

Histone Deacetylase Inhibition as an Anticancer Telomerase-Targeting Strategy

Ruman Rahman
Children's Brain Tumour Research Centre, Medical School,
University of Nottingham, UK

Abstract

Aberrant epigenetic regulation of gene expression contributes to tumour initiation and progression. Studies from a plethora of haematologic and solid tumours support the use of histone deacetylase inhibitors (HDACi) as potent anticancer agents. The mechanism(s) of HDACi-induced cancer cell phenotypes are complex and incompletely elucidated. This chapter will discuss erroneous epigenetic regulation of hTERT transcription in cancer cells and propose that alleviation of an improper acetylation-deacetylation balance at the hTERT promoter, is one mode by which HDACi induces anticancer effects. The chapter will conclude with some pertinent questions and future perspectives arising from the recent impetus in HDACi studies, with particular attention to the cancer stem cell therapeutic paradigm.

Introduction

One of the recurring mechanisms underlying the immortality of cancer cells is the repression of genes encoding tumour suppressors and apoptotic factors; meanwhile genes promoting cell division are up-regulated. These phenomena may be explained in part by the homeostatic maintenance of both histone and non-histone acetylation status. In particular, histone deacetylases (HDACs) are known to regulate gene transcription and oncogenesis through remodelling of chromatin structure, canonically resulting in the repression of target genes [1]. Inhibition of HDACs increases acetylation of histone and non-histone proteins, leading to an increase in transcriptionally active, open chromatin. This conformational change can lead to restoration of transcriptionally silenced tumour suppressor pathways and re-expression of proteins that repress tumour-propagating genes. HDACs have therefore been linked conceptually and mechanistically to the pathogenesis of cancer and are currently under investigation as viable anticancer targets. Small molecule HDAC inhibitors (HDACi) have achieved significant biological effects in preclinical cancer models including glioblastoma multiforme [2], medulloblastoma [3], pancreatic cancer, breast cancer, melanoma and leukaemia [4]. To a variable extent, HDACi induce growth arrest, differentiation or apoptosis *in vitro* and *in vivo* [5, 6]. In some cases, growth arrest is induced at low doses and apoptosis induced at high doses, whereas in other cases growth arrest precedes apoptosis [7].The mechanisms of HDACi-induced cancer cell phenotype(s) are complex and incompletely elucidated. This chapter will review the clinical implications of one such mechanism - HDACi-based telomerase inhibition. Evidence supporting HDACi as a mode of abrogating telomerase activity in tumour cells will be discussed and a hypothesis presented pertaining to the mechanism of action in this context. The chapter will conclude with some pertinent questions and future perspectives arising from the recent impetus in HDACi studies.

Histone (Lysine Tail) Deacetylases and Cancer Therapy

The modification of chromatin is a central tenet of the regulation of gene expression in eukaryotic cells. The covalent modification of core histones by methylation, phosphorylation and acetylation forms a genome-wide combinatorial "code" which controls transcriptional activity of gene sequences wrapped in histone octamers [8]. Aberrant expression of epigenetic regulators of gene expression contributes to tumour initiation and progression. Considerable recent research has focused on the potential of reversing alterations in chromatin structure. Particularly, the acetylation-deacetylation cycle governed by histone acetyltransferases (HATs) and histone deacetylases (HDACs) respectively, may be the basis of transcriptional initiation and repression that allows a rapid response to environmental stimuli [9].

The mammalian histone deacetylase family comprises 18 enzymes grouped into four classes, based on homology to yeast deacetylase proteins. Class I HDACs (HDAC1, -2, -3 and -8), homologues of the yeast Rpd3 protein, are ubiquitously expressed and with the exception of HDAC3, are localized primarily in the nucleus. HDAC4, HDAC5, HDAC7 and HDAC9 belong to class II and have homology to yeast HDA1. HDAC6 and HDAC10 contain two catalytic sites and are classified as IIa, whereas HDAC11 has conserved residues in its catalytic centre that are shared by class I and II deacetylases, and is therefore placed in class IV. Class I, II and IV deacetylases contain zinc in their catalytic site and are inhibited by various HDACi [10, 11]. Class III HDACs include sirtuins, have homology to yeast Sir2 protein and have an absolute requirement for NAD^+. They do not contain zinc in the catalytic site and are therefore not inhibited by conventional HDACi [12, 13].

Since histones were the first identified targets of deacetylases, these enzymes were termed histone deacetylases. It is well established that HDACs catalyze the deacetylation of α-acetyl lysine that resides within the NH_2-terminal tail of core histones. Importantly, HDACs are not exclusively targeted towards histones. Phylogenetic analysis indicates that the evolution of HDACs preceded the evolution of histones, thereby suggesting that primary HDAC targets may not be histones [14]. To date, more than 50 non-histone proteins have been identified that are context-dependent substrates for one or other HDAC [15-17]. Among these non-histone targets are transcription factors, hormone receptors, signal transducers and molecular chaperones [7]. These include p53, MyoD, E2F-1, Stat3, CDK9, SP1 and c-Myc [11]. HDACs may thus more accurately be referred to as "lysine tail deacetylases" [18]. Regulation of acetylation signatures of such proteins (collectively termed the "acetylome") modulates a plethora of cellular events such as proliferation, cell survival and apoptosis [18].

Aberrant expression of HDACs has been observed in various tumour types [19-22] implicating an altered balance of protein acetylation as contributing to cell transformation and cancer initiation/progression. A crude view of target identification for anticancer drug development may suggest HDACs are not suitable targets, as inhibiting HDACs is likely to interfere with critical cellular functions. However, recent investigations prompt a refined view, in which re-expression of non-histone proteins as a result of hyperacetylation of adjacent chromatin, is considered one of the most viable and promising approaches in cancer therapy using HDACi [3, 23-26] Moreover it has been shown that HDACi equally sensitize cancer cells to the cytotoxic effects of other chemotherapeutic agents [27-30] and display selective toxicity against tumour cells compared to untransformed cells [5, 31]. For example, we have recently exposed rat ependymal tissue *ex vivo* to the HDACi trichostatin A (TSA) and observed no adverse effect on ependymal cilia function (Rahman R, manuscript in preparation). The sensitivity of tumour cells and relative resistance of normal cells to HDACi may reflect the multiple defects that render cancer cells less likely than normal cells to compensate for inhibition of one or more pro-survival factors or activation of pro-death pathways. This is exemplified by HDACi triggering a G2 checkpoint arrest in normal cells but which is defective in cancer cells [32]. Indeed the plurality of HDACi target proteins may be important for the efficacy of these agents against a broad spectrum of haematologic and solid tumours.

At least 12 HDACi are currently undergoing clinical trials (spanning over 100 clinical trials) as monotherapy or in combination therapy, for patients with lung, breast, pancreatic, renal and bladder cancers, melanoma, glioblastoma, leukaemias, lymphomas and multiple myelomas [15, 33]. The most widely used HDACi compounds represent one of two classes, the hydroxamic acids (such as SAHA (vorinostat) and TSA) and aliphatic acids (such as valproic acid and sodium butyrate). The latter class are generally weaker HDACi than the hydroxamate acids. From available results, one can conclude that HDACi have activity in haematologic and solid tumours at doses tolerated well by patients. Vorinostat is the first HDACi approved by the Food and Drug Administration (FDA) for the treatment of cutaneous T-cell lymphoma [34]. Elucidation of downstream anticancer pathways activated by HDACi may enable the development of more effective therapeutic strategies with these agents and may help refine synergistic treatment modalities.

Mechanism of HDACi-Induced Cell Death

It is probable that several mechanisms contribute to HDACi-induced anticancer effects, given the multitude of cellular effects triggered by HDACi. The mode of action of a particular HDACi may also be dependent on the context of the cell type or tumour in question and it is possible that more than one mechanism of action occurs during treatment of the same tumour (or cells). A hallmark feature of HDACi is the induction of p21-mediated cell cycle arrest, primarily in the G1 phase [35]. HDACi investigated to date also activates either one or both of the two major apoptotic pathways: the 'extrinsic' death receptor pathway and the 'intrinsic' mitochondrial pathway. Apoptosis stimulated by HDACi is associated with increased expression of pro-apoptotic genes and decreased expression of anti-apoptotic genes, thus shifting the balance towards cell death [36-39]. HDACi also enhances the efficacy of many pro-apoptotic conventional anticancer agents. Consistent with this, overexpression of the anti-apoptotic proteins Bcl-2 and Bcl-X_L block HDACi-mediated apoptosis [40].

Another significant event associated with HDACi is reactive oxygen species (ROS) generation [41, 42]. Although the source of ROS generation remains unclear, these studies suggest that oxidative stress may play an important role in tumour death induced by HDACi. Indeed, treatment with anti-oxidants reduces HDACi potency [42].

In addition to promoting apoptotic cell death, vorinostat and several other HDACi have been reported to inhibit angiogenesis, a key process during tumour progression and metastases. Specifically several pro-angiogenic factors including vascular endothelial growth factor (VEGF) and hypoxia inducible factor-1α (HIF-1α) have been observed to decrease following treatment with HDACi [43, 44]. Conversely overexpression of HDAC1 under hypoxia resulted in reduced levels of p53 and von Hippel-Lindau tumour suppressor genes and stimulated angiogenesis in human endothelial cells [45].

At present, the majority of HDACi *in vitro* and *in vivo* studies have been proof-of-concept and largely descriptive. The temporal order of events upon HDACi exposure, from the alleviation of deacetylase-mediated gene repression (via hyperacetylation) to cellular growth arrest and/or apoptosis is unclear. The precise non-histone molecular targets for each tumour type shown to be amenable for HDACi anticancer strategies remain to be identified in most cases. True genomic substrates which are repressed by particular HDACs and from which transcription can re-initiate upon HDACi-mediated hyperacetylation, must be distinguished from consequential gene expression changes and molecular alterations. Such knowledge will help refine development of novel, more potent HDACi that are selective for tumour cells.

Chromatin Code of hTERT Gene Regulation

Transcriptional regulation of the human telomerase reverse transcriptase (hTERT) gene, encoding the catalytic component of the human telomerase holoenzyme, plays a critical role in the activation of the enzyme during cellular immortalization and tumourigenesis. hTERT confers an immortal phenotype by catalyzing the *de novo* synthesis of telomere repeats at chromosomal termini, using an intrinsic RNA template (hTR). Whereas hTR is expressed constitutively in most human tissues, hTERT is only transiently expressed during S-phase of the cell cycle in most somatic tissues [46]. Thus in normal human somatic cells telomerase activity is tightly regulated by the repression of the hTERT gene. That telomerase is reactivated in ~85% of haematological and solid tumours renders the enzyme as almost a ubiquitous therapeutic target [47]. Although regulation of telomerase activity in mammalian cells is mulitfactorial [48], transcriptional regulation of hTERT appears to be the primary mode of control. Recent findings have provided mechanistic insights into how oncogenic activation as well as tumour de-repression, often due to inactivation of tumour suprressors, stimulates transcription from the hTERT promoter. In tumour cells telomerase activity is closely correlated with the repression and de-repression of hTERT [49]. These findings strongly suggest that de-repression of the hTERT promoter might be an important mechanism leading to the activation of hTERT and thereby of telomerase activity in cells during tumourigenesis. Several studies have indicated that the c-Myc oncogene, de-regulated in several tumours, can contribute to the transcriptional activation of the hTERT gene in tumour cells [50-52].

Sequence analysis has revealed that the hTERT promoter is highly GC-rich, lacks TATA and CAAT boxes, but contains binding sites for a number of transcription factors including the Myc/Max/Mad binding site (E-box) [53]. E-boxes contain the CACGTG motif, recognised by the Myc/Max/Mad network of basic helix-loop-helix/leucine zipper transcription factors. Max can both homodimerize and heterodimerize with Myc and Mad proteins, resulting in transcriptional activation (Myc/Max) or transcriptional repression (Mad/Max) of the hTERT promoter [54]. Myc and Mad antagonize each other and therefore expression levels of the two proteins appear to be inversely regulated. The c-Myc proto-oncogene is activated in proliferating and many neoplastic cells, whereas Mad is primarily expressed in differentiating and resting cells [55]. Two E-boxes upstream of the hTERT start codon have been identified to date that mediate Myc/Max binding and transactivation [56, 57]. Expectedly, these E-boxes also mediate repression of hTERT transcription by Mad [58]. Switching from Myc/Max to Mad/Max occupancy at the hTERT promoter has been observed during differentiation of promyelocytic leukaemic HL60 cells, accompanied by hTERT downregulation [59]. Conversely a reverse switch occurs upon transformation of WI-38 foetal lung fibroblasts with associated acquiring of telomerase activity [60]. The antagonism between Myc and Mad is therefore a crucial determinant of hTERT expression and telomerase activity.

The identified cis-elements at the hTERT promoter also include several Sp1 binding sites at the proximal region which have been shown to be important for hTERT transcriptional activation [50, 57, 61]. Two canonical and three degenerate binding sites for the transcription factor Sp1 are localized within 110bp upstream of the hTERT transcription initiation site. All identified sites interact with Sp1 and mutations at each site reduced hTERT activity [57, 61]. In certain contexts, c-Myc also cooperates with Sp1 to induce hTERT transcription [61]. However, serial deletion analysis of the hTERT promoter in normal human fibroblasts identified critical repressive elements that formed Sp1 (and the related Sp3) DNA-protein complexes in nuclear extracts. A number of groups concurrently showed that these elements were HDAC1 and 2, recruited to the hTERT promoter by Sp1 and Sp3 to thereby inhibit the hTERT transcription [49, 62]. Overexpression of dominant-negative Sp1 and Sp3 mutants, which contained mainly the HDAC2-binding domain, relieved the HDAC-mediated repression of the hTERT promoter. In addition, repression of the hTERT promoter by Mad protein also requires HDAC which may be recruited by Mad repressor complexes bound to E-boxes within the hTERT promoter [53]. Experiments in a wide variety of normal human somatic cells treated with the HDACi TSA, suggest that histone deacetylation is essential to the transcriptional regulation of the hTERT gene. TSA induces hyperacetylation of histones at the hTERT proximal promoter and directly transactivates the hTERT gene in normal telomerase-negative/low cells. TSA-mediated activation of the hTERT promoter is abolished by the mutation of Sp1 sites suggesting the effect of TSA is regulated through Sp1 motifs. Up-regulation of hTERT promoter activity results in telomerase activity levels comparable with that in human tumour cells [49, 62]. Taken together, these results suggest that Sp1, Sp3 and Mad proteins associate with the hTERT promoter, recruiting HDACs for the localized deacetylation of nucleosomal histones and transcriptional silencing of the hTERT gene in normal human somatic cells. Such repression mediated by HDACs can be either dependent (Sp1 and Sp3) or independent (Mad) of E-boxes on the hTERT promoter. Thus, HDAC-mediated repression could be the major, universal transcriptional repression mechanism of the hTERT gene in normal human somatic cells.

Transformimg growth factor β (TGF-β) is known to carry out tumour suppressor functions in epithelial and lymphoid cells, from which most human cancers arise. Levels of TGF-β are inversely correlated to the levels of telomerase activity [63] and TGF-β is a potent telomerase-inhibiting growth factor in cells cultured under physiologically relevant conditions [64]. As Smad proteins (Smad1-8) are downstream effectors of TGF-β signalling and Smad3 specifically mediates TGF-β action, the inhibition of hTERT by TGF-β suggests a cis action of the TGF-β signalling molecule Smad3 on the hTERT promoter directly [65]. While the mechanism(s) underlying hTERT gene repression induced by TGF-β remain unclear, it is possible that this repression is at least in part, mediated by TGF-β-suppression of the c-Myc proto-oncogene. The c-Myc gene is downregulated through the TGF-β/Smad pathway [66, 67] and may account for how TGF-β stimulation can suppress hTERT expression and thereby exert its tumour suppressive functions [68, 69]. Recent manipulation of intracellular Smad3 gene expression has revealed that Smad3 interacts with a specific site on the hTERT promoter in response to TGF-β stimulation *in vitro* and in intact cells, leading to a significant inhibition of hTERT transcription. Thus, Smad3 may constitute a negative regulatory system to balance c-Myc transcriptional activation of hTERT. Analysis of the hTERT gene promoter reveals multiple putative binding sites for Smad3, adjacent to E-box of the c-Myc binding site [65]. Other studies reveal protein-protein interactions between Smad3 and c-Myc [70]. Binding to c-Myc directly may impact upon c-Myc regulation of hTERT. Whether Smad3 exerts its repressive function on hTERT gene transcription via protein-protein interactions with c-Myc or direct binding to hTERT promoter DNA, is unclear at present and may not be mutually exclusive.

Consistent with findings of HDAC recruitment to sites of hTERT transcription initiation, TGF-β inhibits hTERT gene expression in a HDAC-dependent manner. While class I HDAC1 did not exhibit any significant effect on hTERT promoter activity, the two class II HDAC4 and HDAC5 significantly repressed hTERT expression. Increasing concentrations of TSA completely reversed the inhibitory effect of TGF-β on the hTERT promoter [71]. TGF-β still potently repressed transcription of the hTERT promoter when both c-Myc binding sites were removed, suggesting that TGF-β utilizes distinct mechanisms to repress telomerase activity in a cell-specific manner [71].

Targeting Telomerase (Regulators) in Cancer Cells Using HDACi

The data reviewed thus far is consistent with the paradigm that the endogenous hTERT promoter in normal human somatic cells that exhibit no/low telomerase activity is repressed by mechanisms involving local chromatin configuration, such as those governed by acetylation homeostasis. De-repression of hTERT transcription (via loss of tumour suppressor function) is therefore likely to be involved in cellular immortalization as a result of telomerase activation in cancer cells. This view does not conflict with the canonical description of hTERT being aberrantly activated through the action of oncoproteins on the hTERT promoter, as it is plausible that both phenomena may occur sequentially in the same cell.

Human chromosomes terminate in specialized nucleoprotein structures termed telomeres, which consist of a duplex array of hexameric TTAGGG DNA repeats (5-15kb) with a 3' single-strand overhang that invades the double-stranded telomeric region to form a protective cap (t-loop) [72]. A complex formed by six telomere-specific proteins associates with this sequence and protects chromosome ends, referred to as the shelterin complex [73]. Disruption of telomere structure by erosion of telomere repeats (cellular ageing) or displacement of t-loop structure activates a signal transduction program that closely resembles the cellular responses upon DNA damage. Telomere dysfunction subsequently induces an irreversible proliferation arrest know as replicative senescence and/or apoptosis [74, 75]. As most tumour cells use telomerase-mediated telomere maintenance to confer cellular immortalization, likely in considerable part due to de-repression of hTERT transcription, chemotherapeutics directed at suppression of hTERT transcription are an attractive anticancer strategy. Many therapeutic strategies that target telomerase using oligonucleotides or small molecule inhibitors, are reliant on telomere shortening (at least on a subset of telomeres) beyond a critical minimum required for a protective t-loop structure. Hence, the efficacy of these approaches are dependent on initial telomere length in a given tumour cell [76]. Strategies that induce rapid telomere dysfunction and chromosomal instability are therefore attractive propositions.

A number of recent studies have assessed the effects of treating tumour cells with HDACi, based on the rationale that the acetylation-deacetylation balance in neoplastic cells is imbalanced relative to normal cells, with respect to hTERT transcriptional regulators. Thus HDACi-mediated alleviation of aberrant deacetylation may in theory compromise telomerase activity (although HDACi will lead to hyperacetylation of histone and non-histone proteins across the genome). Shigeru Kohno and colleagues initially reported that HDACi (TSA and sodium butyrate) suppressed hTERT mRNA expression and telomerase activity in human prostate cancer cells. Inhibition of hTERT expression preceded the suppression of cell proliferation, suggesting that HDACi-mediated effects on the hTERT promoter are early upstream events of HDACi exposure. Neither TSA nor sodium butyrate however, altered the expression of c-Myc or p21 at the stage of HDACi treatment at which hTERT mRNA inhibition was observed. It is unclear whether HDACi repress hTERT transcription directly in this context, or whether the expression of modulating factors other than c-Myc or p21 is suppressed upon HDACi [77]. HDACi treatment of human leukaemic cells similarly resulted in inhibition of telomerase activity and hTERT gene expression, with concurrent induction of apoptosis in a dose-dependent manner using TSA [78]. Telomerase inhibition following HDACi treatment has also been observed in cervical cancer cells and liver cells [79, 80].

We and others have demonstrated inhibition of telomerase activity and hTERT gene expression in brain tumour cells treated with TSA [2], (Rahman R manuscript in preparation) (Figure 1). In both cases, exposure to TSA resulted in apoptotic induction in a dose-dependent manner and was associated with p21 upregulation. Intriguingly, two recent studies observe that HDACi-mediated activation of p21 coincides with dissociation of c-Myc and HDAC1 within the p21 promoter [81, 82]. We further demonstrated that telomerase inhibition was specifically due to HDACi activity rather than a consequence of caspase-3-dependent apoptosis; exposure to the DNA damage agent, etoposide, resulted in caspase-3-dependent apoptosis without concomitant abrogation of telomerase activity. Furthermore we observed no toxicity in an *ex vivo* normal rat ependymal model with respect to ependymal cilia function (Rahman R manuscript in preparation). It is unclear from any study to date as to why HDACi induces hTERT transcription in normal somatic cells, whereas suppresses hTERT transcription in cancer cells. It is plausible that the alleviation of hTERT and/or c-Myc repressor protein silencing overrides hyperacetylation at the hTERT promoter itself.

Although most studies to date have reported on non-histone targets modulating the cellular response to HDACi, a recent finding demonstrates that histones directly may be attractive targets for directed HDACi therapy. The human SIRT6 protein is an NAD^+ - dependent histone H3 lysine 9 (H3K9) deacetylase that modulates telomeric chromatin by associating specifically with telomeres. SIRT6 depletion leads to telomere dysfunction with end-to-end chromosomal fusions and premature cellular senescence [83]. If SIRT6 contributes to the propagation of a specialized chromatin state at human telomeres, then it is possible that novel HDACi may be developed to target SIRT6, leading to telomere structure disruption. It will be important to elucidate whether there are other HDACs that modulate telomeric chromatin specifically.

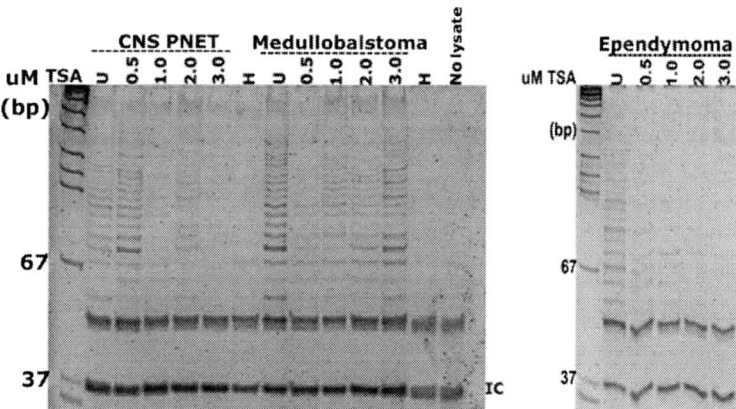

Figure 1. Telomerase inhibition upon HDACi in brain tumour cells. The telomere repeat amplification protocol (TRAP) was used to determine TSA-induced (0.5-3.0μM) effects on telomerase activity on paediatric brain tumour cells of neuroectodermal (CNS PNET, medulloblastoma) and glial (ependymoma) origin. TSA exposure (48h) results in marked inhibition of telomerase activity in all cell lines. *U, untreated cells; H, telomerase heat-inactivated control; No lysate, buffer-only negative control.* (Rahman R et al, manuscript in preparation).

Summary and Perspectives

Development of targeted cancer therapy is classically based using gene- or pathway-specific methodology. In the case of HDACi, the use of the drug preceded the molecular description and characterization of the target. It is now clear that HDACi exert a multitude of cell-intrinsic anticancer mechanisms, such as cell cycle arrest, apoptosis and differentiation. Furthermore it is evident that HDACi act both in an epigenetic and non-epigenetic manner, targeting lysine tails of histone and non-histone proteins respectively. Indeed, it unclear as to whether targeting of epigenetic mechanisms and chromatin remodelling, or modulation of the functional activity of cytoplasmic proteins and transcription factors (or both mechanisms), contribute most to anti-cancer activity of these compounds.

This chapter has focused on one putative location for HDACi action – the hTERT promoter. Numerous studies have shown that hTERT gene expression and consequential abrogation of telomerase activity is an early upstream event following HDACi, proceeded by cell phenotypic effects such as proliferation arrest and apoptosis. It is likely that an improper balance of acetylation-deacetylation status in cancerous cells results in aberrant activation of hTERT transcriptional activators such as c-Myc and aberrant repression of hTERT transcriptional silencers such as Mad and Smad3. Thus, I propose a hypothetical model whereby HDACi proximally results in hyperacetylation of lysine tails on Mad or Smad3 (or alternatively an unidentified repressor specific to c-Myc), resulting in alleviation of Mad or Smad3 repression. The subsequent reactivation of hTERT repressor proteins would ultimately result in a shift of balance regarding acetylation status of hTERT modulators, leading to inhibition of hTERT gene expression and a negative effect on cell viability and growth as a result (Figure 2). Of course, HDACi must affect the genome-wide acetylation signature; however as telomerase-mediated telomere maintenance is pivotal for cellular immortalization, HDACi effects on the hTERT promoter are likely to represent a key upstream event. Further studies are required to refine our knowledge of HDACi targets such as hTERT. It is not known whether pan-HDACi that target several HDACs or novel HDACi that are specific to one HDAC or one HDAC class, will emerge as the most potent HDACi for next generation cancer therapy trials.

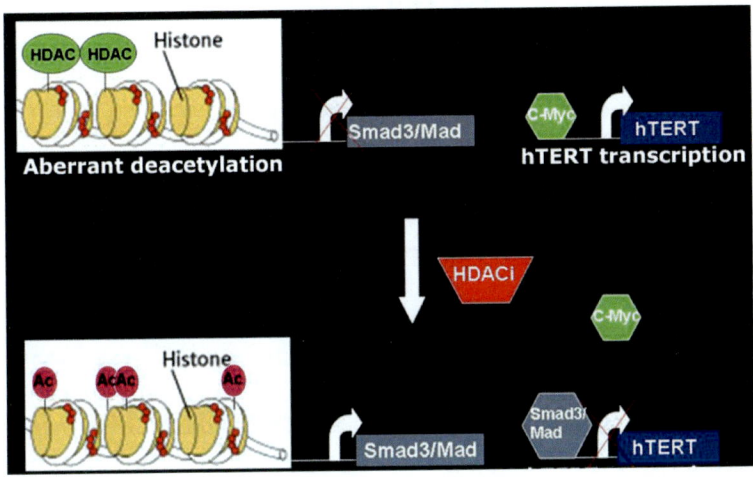

Figure 2. Proposed model of HDACi-induced hTERT repression. (Top) hTERT transcriptional repressors such as Smad3 and Mad may be aberrantly silenced in neoplastic cells due to aberrant deacetylation of lysine tails in adjacent chromatin. This results in a failure to repress hTERT transcription initiation from c-Myc (i.e. derepression of the hTERT promoter). (Bottom) Removal of histone deacetylases through the use of HDACi, results in hyperacetylation of lysine tails and alleviation of Smad3/Mad silencing. Smad3 and Mad may then conduct the canonical roles of repression of hTERT transcription, via displacement of c-Myc in the case of Mad. This latter scenario may recapitulate the status of hTERT regulation in normal human somatic cells.

Some pertinent questions bring our current knowledge of HDACs and HDACi in line with current dogma in molecular cancer biology: is epigenetics or non-epigenetics the critical mechanism underlying the anticancer effects of HDACi? what is the status of telomerase activity and telomere length in tumour stem/progenitor cells and can HDACi exert anticancer effects in these tumour-initiating subpopulations? how does the microenvironment of the tumour (including regions of hypoxia) influence HDACi efficacy and vice versa? what are the best combinatorial strategies to include HDACi in prospective clinical trials?

Elucidation of such questions will require a consilient approach that encourages collaboration between students of cancer biology, telomere/telomerase biology, epigenetics and cancer therapeutics.

References

[1] Richon, VM; O'Brien, JP. Histone deacetylase inhibitors: a new class of potential therapeutic agents for cancer treatment. *Clin Cancer Res.*, 2002, Mar, 8(3), 662-4.

[2] Khaw, AK; Silasudjana, M; Banerjee, B; Suzuki, M; Baskar, R; Hande, MP. Inhibition of telomerase activity and human telomerase reverse transcriptase gene expression by histone deacetylase inhibitor in human brain cancer cells. *Mutat Res.*, 2007, Dec 1, 625(1-2), 134-44.

[3] Li, XN; Shu, Q; Su, JM; Perlaky, L; Blaney, SM; Lau, CC. Valproic acid induces growth arrest, apoptosis, and senescence in medulloblastomas by increasing histone hyperacetylation and regulating expression of p21Cip1, CDK4, and CMYC. *Mol Cancer Ther.*, 2005, Dec, 4(12), 1912-22.

[4] Gargiulo, G; Minucci, S. Epigenomic profiling of cancer cells. *Int J Biochem Cell Biol.*, 2009, Jan, 41(1), 127-35.

[5] Johnstone, RW; Ruefli, AA; Lowe, SW. Apoptosis: a link between cancer genetics and chemotherapy. *Cell.*, 2002, Jan 25, 108(2), 153-64.

[6] Marks, PA; Rifkind, RA; Richon, VM; Breslow, R. Inhibitors of histone deacetylase are potentially effective anticancer agents. *Clin Cancer Res.*, 2001, Apr, 7(4), 759-60.

[7] Minucci, S; Pelicci, PG. Histone deacetylase inhibitors and the promise of epigenetic (and more) treatments for cancer. *Nat Rev Cancer*, 2006, Jan, 6(1), 38-51.

[8] Jenuwein, T; Allis, CD. Translating the histone code. *Science*, 2001, Aug 10, 293(5532), 1074-80.

[9] Metivier, R; Penot, G; Hubner, MR; Reid, G; Brand, H; Kos, M; et al. Estrogen receptor-alpha directs ordered, cyclical, and combinatorial recruitment of cofactors on a natural target promoter. *Cell*, 2003, Dec 12, 115(6), 751-63.

[10] Buchwald, M; Kramer, OH; Heinzel, T. HDACi--targets beyond chromatin. *Cancer Lett*, 2009, Aug 8, 280(2), 160-7.

[11] Dokmanovic, M; Clarke, C; Marks, PA. Histone deacetylase inhibitors: overview and perspectives. *Mol Cancer Res.*, 2007, Oct, 5(10), 981-9.

[12] Bolden, JE; Peart, MJ; Johnstone, RW. Anticancer activities of histone deacetylase inhibitors. *Nat Rev Drug Discov*, 2006, Sep, 5(9), 769-84.

[13] Marks, P; Rifkind, RA; Richon, VM; Breslow, R; Miller, T; Kelly, WK. Histone deacetylases and cancer: causes and therapies. *Nat Rev Cancer*, 2001, Dec, 1(3), 194-202.

[14] Gregoretti, IV; Lee, YM; Goodson, HV. Molecular evolution of the histone deacetylase family: functional implications of phylogenetic analysis. *J Mol Biol.*, 2004, Apr 16, 338(1), 17-31.
[15] Dokmanovic, M; Marks, PA. Prospects: histone deacetylase inhibitors. *J Cell Biochem*, 2005, Oct 1, 96(2), 293-304.
[16] Marks, PA; Breslow, R. Dimethyl sulfoxide to vorinostat: development of this histone deacetylase inhibitor as an anticancer drug. *Nat Biotechnol*, 2007, Jan, 25(1), 84-90.
[17] Rosato, RR; Grant, S. Histone deacetylase inhibitors: insights into mechanisms of lethality. *Expert Opin Ther Targets*, 2005, Aug, 9(4), 809-24.
[18] Xu, WS; Parmigiani, RB; Marks, PA. Histone deacetylase inhibitors: molecular mechanisms of action. *Oncogene*, 2007, Aug 13, 26(37), 5541-52.
[19] Sakuma, T; Uzawa, K; Onda, T; Shiiba, M; Yokoe, H; Shibahara, T; et al. Aberrant expression of histone deacetylase 6 in oral squamous cell carcinoma. *Int J Oncol*, 2006, Jul, 29(1), 117-24.
[20] Wilson, AJ; Byun, DS; Popova, N; Murray, LB; L'Italien, K; Sowa, Y; et al. Histone deacetylase 3 (HDAC3) and other class I HDACs regulate colon cell maturation and p21 expression and are deregulated in human colon cancer. *J Biol Chem.*, 2006, May 12, 281(19), 13548-58.
[21] Zhang, Z; Chen, CQ; Manev, H. DNA methylation as an epigenetic regulator of neural 5-lipoxygenase expression: evidence in human NT2 and NT2-N cells. *J Neurochem*, 2004, Mar, 88(6), 1424-30.
[22] Zhu, P; Martin, E; Mengwasser, J; Schlag, P; Janssen, KP; Gottlicher, M. Induction of HDAC2 expression upon loss of APC in colorectal tumorigenesis. *Cancer Cell*, 2004, May, 5(5), 455-63.
[23] Kim, HR; Kim, EJ; Yang, SH; Jeong, ET; Park, C; Lee, JH; et al. Trichostatin A induces apoptosis in lung cancer cells via simultaneous activation of the death receptor-mediated and mitochondrial pathway? *Exp Mol Med*, 2006, Dec 31, 38(6), 616-24.
[24] Qu, W; Kang, YD; Zhou, MS; Fu, LL; Hua, ZH; Wang, LM. Experimental study on inhibitory effects of histone deacetylase inhibitor MS-275 and TSA on bladder cancer cells. *Urol Oncol*, 2009, Jan 30.
[25] Wu, Y; Guo, SW. Histone deacetylase inhibitors trichostatin A and valproic acid induce cell cycle arrest and p21 expression in immortalized human endometrial stromal cells. *Eur J Obstet Gynecol Reprod Biol.*, 2008, Apr, 137(2), 198-203.
[26] Yin, D; Ong, JM; Hu, J; Desmond, JC; Kawamata, N; Konda, BM; et al.

Suberoylanilide hydroxamic acid, a histone deacetylase inhibitor: effects on gene expression and growth of glioma cells in vitro and in vivo. *Clin Cancer Res.*, 2007, Feb 1, 13(3), 1045-52.

[27] Camphausen, K; Cerna, D; Scott, T; Sproull, M; Burgan, WE; Cerra, MA; et al. Enhancement of in vitro and in vivo tumor cell radiosensitivity by valproic acid. *Int J Cancer*, 2005, Apr 10, 114(3), 380-6.

[28] Dowdy, SC; Jiang, S; Zhou, XC; Hou, X; Jin, F; Podratz, KC; et al. Histone deacetylase inhibitors and paclitaxel cause synergistic effects on apoptosis and microtubule stabilization in papillary serous endometrial cancer cells. *Mol Cancer Ther.*, 2006, Nov, 5(11), 2767-76.

[29] Ecke, I; Petry, F; Rosenberger, A; Tauber, S; Monkemeyer, S; Hess, I; et al. Antitumor effects of a combined 5-aza-2'deoxycytidine and valproic acid treatment on rhabdomyosarcoma and medulloblastoma in Ptch mutant mice. *Cancer Res.*, 2009, Feb 1, 69(3), 887-95.

[30] Kim, IA; Shin, JH; Kim, IH; Kim, JH; Kim, JS; Wu, HG; et al. Histone deacetylase inhibitor-mediated radiosensitization of human cancer cells: class differences and the potential influence of p53. *Clin Cancer Res.*, 2006, Feb 1, 12(3 Pt 1), 940-9.

[31] Gottlicher, M; Minucci, S; Zhu, P; Kramer, OH; Schimpf, A; Giavara, S; et al. Valproic acid defines a novel class of HDAC inhibitors inducing differentiation of transformed cells. *EMBO J*, 2001, Dec 17, 20(24), 6969-78.

[32] Qiu, L; Burgess, A; Fairlie, DP; Leonard, H; Parsons, PG; Gabrielli, BG. Histone deacetylase inhibitors trigger a G2 checkpoint in normal cells that is defective in tumor cells. *Mol Biol Cell.*, 2000, Jun, 11(6), 2069-83.

[33] Rasheed, WK; Johnstone, RW; Prince, HM. Histone deacetylase inhibitors in cancer therapy. *Expert Opin Investig Drugs*, 2007, May, 16(5), 659-78.

[34] Duvic, M; Vu, J. Vorinostat: a new oral histone deacetylase inhibitor approved for cutaneous T-cell lymphoma. *Expert Opin Investig Drugs*, 2007, Jul, 16(7), 1111-20.

[35] Carew, JS; Giles, FJ; Nawrocki, ST. Histone deacetylase inhibitors: mechanisms of cell death and promise in combination cancer therapy. *Cancer Lett*, 2008, Sep 28, 269(1), 7-17.

[36] Imai, T; Adachi, S; Nishijo, K; Ohgushi, M; Okada, M; Yasumi, T; et al. FR901228 induces tumor regression associated with induction of Fas ligand and activation of Fas signaling in human osteosarcoma cells.

Oncogene, 2003, Dec 18, 22(58), 9231-42.
[37] Rosato, RR; Almenara, JA; Dai, Y; Grant, S. Simultaneous activation of the intrinsic and extrinsic pathways by histone deacetylase (HDAC) inhibitors and tumor necrosis factor-related apoptosis-inducing ligand (TRAIL) synergistically induces mitochondrial damage and apoptosis in human leukemia cells. *Mol Cancer Ther.*, 2003, Dec, 2(12), 1273-84.
[38] Sonnemann, J; Hartwig, M; Plath, A; Saravana Kumar, K; Muller, C; Beck, JF. Histone deacetylase inhibitors require caspase activity to induce apoptosis in lung and prostate carcinoma cells. *Cancer Lett*, 2006, Feb 8, 232(2), 148-60.
[39] Sonnemann, J; Kumar, KS; Heesch, S; Muller, C; Hartwig, C; Maass, M; et al. Histone deacetylase inhibitors induce cell death and enhance the susceptibility to ionizing radiation, etoposide, and TRAIL in medulloblastoma cells. *Int J Oncol*, 2006, Mar, 28(3), 755-66.
[40] Ruefli, AA; Ausserlechner, MJ; Bernhard, D; Sutton, VR; Tainton, KM; Kofler, R; et al. The histone deacetylase inhibitor and chemotherapeutic agent suberoylanilide hydroxamic acid (SAHA) induces a cell-death pathway characterized by cleavage of Bid and production of reactive oxygen species. *Proc Natl Acad Sci U S A*, 2001, Sep 11, 98(19), 10833-8.
[41] Butler, LM; Zhou, X; Xu, WS; Scher, HI; Rifkind, RA; Marks, PA; et al. The histone deacetylase inhibitor SAHA arrests cancer cell growth, up-regulates thioredoxin-binding protein-2, and down-regulates thioredoxin. *Proc Natl Acad Sci U S A*, 2002, Sep 3, 99(18), 11700-5.
[42] Ungerstedt, JS; Sowa, Y; Xu, WS; Shao, Y; Dokmanovic, M; Perez, G; et al. Role of thioredoxin in the response of normal and transformed cells to histone deacetylase inhibitors. *Proc Natl Acad Sci U S A*, 2005, Jan 18, 102(3), 673-8.
[43] Deroanne, CF; Bonjean, K; Servotte, S; Devy, L; Colige, A; Clausse, N; et al. Histone deacetylases inhibitors as anti-angiogenic agents altering vascular endothelial growth factor signaling. *Oncogene*, 2002, Jan 17, 21(3), 427-36.
[44] Qian, DZ; Kachhap, SK; Collis, SJ; Verheul, HM; Carducci, MA; Atadja, P; et al. Class II histone deacetylases are associated with VHL-independent regulation of hypoxia-inducible factor 1 alpha. *Cancer Res.*, 2006, Sep 1, 66(17), 8814-21.
[45] Kim, MS; Kwon, HJ; Lee, YM; Baek, JH; Jang, JE; Lee, SW; et al. Histone deacetylases induce angiogenesis by negative regulation of tumor suppressor genes. *Nat Med*, 2001, Apr, 7(4), 437-43.

[46] Masutomi, K; Yu, EY; Khurts, S; Ben-Porath, I; Currier, JL; Metz, GB; et al. Telomerase maintains telomere structure in normal human cells. *Cell*, 2003, Jul 25, 114(2), 241-53.
[47] Kim, NW; Piatyszek, MA; Prowse, KR; Harley, CB; West, MD; Ho, PL; et al. Specific association of human telomerase activity with immortal cells and cancer. *Science*, 1994, Dec 23, 266(5193), 2011-5.
[48] Liu, JP. Studies of the molecular mechanisms in the regulation of telomerase activity. *FASEB J*, 1999, Dec, 13(15), 2091-104.
[49] Won, J; Yim, J; Kim, TK. Sp1 and Sp3 recruit histone deacetylase to repress transcription of human telomerase reverse transcriptase (hTERT) promoter in normal human somatic cells. *J Biol Chem.*, 2002, Oct 11, 277(41), 38230-8.
[50] Horikawa, I; Cable, PL; Afshari, C; Barrett, JC. Cloning and characterization of the promoter region of human telomerase reverse transcriptase gene. *Cancer Res.*, 1999, Feb 15, 59(4), 826-30.
[51] Oh, S; Song, YH; Kim, UJ; Yim, J; Kim, TK. In vivo and in vitro analyses of Myc for differential promoter activities of the human telomerase (hTERT) gene in normal and tumor cells. *Biochem Biophys Res Commun.*, 1999, Sep 24, 263(2), 361-5.
[52] Wu, KJ; Grandori, C; Amacker, M; Simon-Vermot, N; Polack, A; Lingner, J; et al. Direct activation of TERT transcription by c-MYC. *Nat Genet*, 1999, Feb, 21(2), 220-4.
[53] Cong, YS; Bacchetti, S. Histone deacetylation is involved in the transcriptional repression of hTERT in normal human cells. *J Biol Chem*, 2000, Nov 17, 275(46), 35665-8.
[54] Janknecht, R. On the road to immortality: hTERT upregulation in cancer cells. *FEBS Lett*, 2004, Apr 23, 564(1-2), 9-13.
[55] Luscher, B. Function and regulation of the transcription factors of the Myc/Max/Mad network. *Gene*, 2001, Oct 17, 277(1-2), 1-14.
[56] Greenberg, RA; O'Hagan, RC; Deng, H; Xiao, Q; Hann, SR; Adams, RR; et al. Telomerase reverse transcriptase gene is a direct target of c-Myc but is not functionally equivalent in cellular transformation. *Oncogene*, 1999, Feb 4, 18(5), 1219-26.
[57] Takakura, M; Kyo, S; Kanaya, T; Hirano, H; Takeda, J; Yutsudo, M; et al. Cloning of human telomerase catalytic subunit (hTERT) gene promoter and identification of proximal core promoter sequences essential for transcriptional activation in immortalized and cancer cells. *Cancer Res.*, 1999, Feb 1, 59(3), 551-7.
[58] Gunes, C; Lichtsteiner, S; Vasserot, AP; Englert, C. Expression of the

hTERT gene is regulated at the level of transcriptional initiation and repressed by Mad1. *Cancer Res.*, 2000, Apr 15, 60(8), 2116-21.

[59] Xu, D; Popov, N; Hou, M; Wang, Q; Bjorkholm, M; Gruber, A; et al. Switch from Myc/Max to Mad1/Max binding and decrease in histone acetylation at the telomerase reverse transcriptase promoter during differentiation of HL60 cells. *Proc Natl Acad Sci U S A*, 2001, Mar 27, 98(7), 3826-31.

[60] Casillas, MA; Brotherton, SL; Andrews, LG; Ruppert, JM; Tollefsbol, TO. Induction of endogenous telomerase (hTERT) by c-Myc in WI-38 fibroblasts transformed with specific genetic elements. *Gene*, 2003, Oct 16, 316, 57-65.

[61] Kyo, S; Takakura, M; Taira, T; Kanaya, T; Itoh, H; Yutsudo, M; et al. Sp1 cooperates with c-Myc to activate transcription of the human telomerase reverse transcriptase gene (hTERT). *Nucleic Acids Res.*, 2000, Feb 1, 28(3), 669-77.

[62] Hou, M; Wang, X; Popov, N; Zhang, A; Zhao, X; Zhou, R; et al. The histone deacetylase inhibitor trichostatin A derepresses the telomerase reverse transcriptase (hTERT) gene in human cells. *Exp Cell Rei.*, 2002, Mar 10, 274(1), 25-34.

[63] Yang, H; Kyo, S; Takatura, M; Sun, L. Autocrine transforming growth factor beta suppresses telomerase activity and transcription of human telomerase reverse transcriptase in human cancer cells. *Cell Growth Differ.*, 2001, Feb, 12(2), 119-27.

[64] Bayne, S; Liu, JP. Hormones and growth factors regulate telomerase activity in ageing and cancer. *Mol Cell Endocrinol*, 2005, Aug 30, 240(1-2), 11-22.

[65] Li, H; Xu, D; Li, J; Berndt, MC; Liu, JP. Transforming growth factor beta suppresses human telomerase reverse transcriptase (hTERT) by Smad3 interactions with c-Myc and the hTERT gene. *J Biol Chem.*, 2006, Sep 1, 281(35), 25588-600.

[66] Chen, CR; Kang, Y; Siegel, PM; Massague, J. E2F4/5 and p107 as Smad cofactors linking the TGFbeta receptor to c-myc repression. *Cell*, 2002, Jul 12, 110(1), 19-32.

[67] Yagi, K; Furuhashi, M; Aoki, H; Goto, D; Kuwano, H; Sugamura, K; et al. c-myc is a downstream target of the Smad pathway. *J Biol Chem.*, 2002, Jan 4, 277(1), 854-61.

[68] Katakura, Y; Nakata, E; Miura, T; Shirahata, S. Transforming growth factor beta triggers two independent-senescence programs in cancer cells. *Biochem Biophys Res Commun.*, 1999, Feb 5, 255(1), 110-5.

[69] Zhu, X; Kumar, R; Mandal, M; Sharma, N; Sharma, HW; Dhingra, U; et al. Cell cycle-dependent modulation of telomerase activity in tumor cells. *Proc Natl Acad Sci U S A*, 1996, Jun 11, 93(12), 6091-5.
[70] Feng, XH; Liang, YY; Liang, M; Zhai, W; Lin, X. Direct interaction of c-Myc with Smad2 and Smad3 to inhibit TGF-beta-mediated induction of the CDK inhibitor p15(Ink4B). *Mol Cell*, 2002 Jan, 9(1), 133-43.
[71] Lacerte, A; Korah, J; Roy, M; Yang, XJ; Lemay, S; Lebrun, JJ. Transforming growth factor-beta inhibits telomerase through SMAD3 and E2F transcription factors. *Cell Signal*, 2008, Jan, 20(1), 50-9.
[72] Griffith, JD; Comeau, L; Rosenfield, S; Stansel, RM; Bianchi, A; Moss H; et al. Mammalian telomeres end in a large duplex loop. *Cell*, 1999, May 14, 97(4), 503-14.
[73] de Lange, T. Shelterin: the protein complex that shapes and safeguards human telomeres. *Genes Dev*, 2005, Sep 15, 19(18), 2100-10.
[74] Hahn, WC; Meyerson, M. Telomerase activation, cellular immortalization and cancer. *Ann Med*, 2001, Mar, 33(2), 123-9.
[75] Oulton, R; Harrington, L. Telomeres, telomerase, and cancer: life on the edge of genomic stability. *Curr Opin Oncol*, 2000, Jan, 12(1), 74-81.
[76] Harley, CB. Telomerase and cancer therapeutics. *Nat Rev Cancer*, 2008, Mar, 8(3), 167-79.
[77] Suenaga, M; Soda, H; Oka, M; Yamaguchi, A; Nakatomi, K; Shiozawa, K; et al. Histone deacetylase inhibitors suppress telomerase reverse transcriptase mRNA expression in prostate cancer cells. *Int J Cancer*, 2002, Feb 10, 97(5), 621-5.
[78] Woo, HJ; Lee, SJ; Choi, BT; Park, YM; Choi, YH. Induction of apoptosis and inhibition of telomerase activity by trichostatin A, a histone deacetylase inhibitor, in human leukemic U937 cells. *Exp Mol Pathol*, 2007, Feb, 82(1), 77-84.
[79] Nakamura, M; Saito, H, Ebinuma, H; Wakabayashi, K; Saito, Y; Takagi T; et al. Reduction of telomerase activity in human liver cancer cells by a histone deacetylase inhibitor. *J Cell Physiol*, 2001, Jun, 187(3), 392-401.
[80] Wu, P; Meng, L; Wang, H; Zhou, J; Xu, G; Wang, S; et al. Role of hTERT in apoptosis of cervical cancer induced by histone deacetylase inhibitor. *Biochem Biophys Res Commun.*, 2005, Sep 16, 335(1), 36-44.
[81] Gui, CY; Ngo, L; Xu, WS; Richon, VM; Marks, PA. Histone deacetylase (HDAC) inhibitor activation of p21WAF1 involves changes in promoter-associated proteins, including HDAC1. *Proc Natl Acad Sci U S A*, 2004, Feb 3, 101(5), 1241-6.

[82] Li, H; Wu, X. Histone deacetylase inhibitor, Trichostatin A, activates p21WAF1/CIP1 expression through downregulation of c-myc and release of the repression of c-myc from the promoter in human cervical cancer cells. *Biochem Biophys Res Commun*, 2004, Nov 12, 324(2), 860-7.

[83] Michishita, E; McCord, RA; Berber, E; Kioi, M; Padilla-Nash, H; Damian, M; et al. SIRT6 is a histone H3 lysine 9 deacetylase that modulates telomeric chromatin. *Nature*, 2008, Mar 27, 452(7186), 492-6.

In: Telomerase: Composition, Functions ...   ISBN: 978-1-61668-957-5
Editor: Aiden N. Gagnon, pp. 81-94   © 2010 Nova Science Publishers, Inc.

Chapter IV

Telomerase Role in Pituitary Adenomas

A. Ortiz-Plata
Laboratory of Experimental Neuropathology,
National Institute of Neurology and Neurosurgery, Mexico City, Mexico

Abstract

Telomeres are DNA-protein structures located at ends of each chromosome, which function is to protect the ends of the DNA double helix. Telomeres are implicated in complex biological processes such as regulation of gene expression, cellular senescence, and tumorigenesis. Telomerase is a ribonucleoprotein enzyme that catalyzes the cellular synthesis of telomeric DNA during cellular division, resulting in maintenance of telomere length and increased proliferative potential. Its activity is directly related with the expression of the catalytic component hTERT (human telomerase reverse transcriptase). Several studies suggest that the telomerase may play an important role in the diagnosis and prognosis of cancer because its expression strongly correlates with the potential tumor progression. Ninety percent of human cancers on different organs have shown high telomerasa activity. In cerebral tumors the telomerase activity has been observed on astrocitomas, multiform glioblastomas, meningiomas, oligodendrogliomas and metastatic cerebral tumors, and these results suggest that telomerase activity may be an important marker of brain tumor malignancy. On the other hand, the telomerase activity has not been detected on benign tumors such as the hemangioblastomas and schwannomas, and its expression and the role it plays on the hypophyseal adenomas has not been cleared yet. The hypophysis adenomas represent 10-15 % of intercraneal tumors. Histologically, they are mostly benign tumors that do not show cellular pleomorfism or mitosis figures. However, in some cases they show rapid growth and can spread to nearby structures, have recurrence, be clinically aggressive and even malignize. In spite of the clinical - pathological analysis and the evaluations of the expression of different molecules, aggressiveness and biological behavior of these "malignant" adenomas has not been found. The expression of the catalytic fraction of the telomerase (hTERT) in this type of tumors has been few analyzed. In some reports the hTERT activity has not been detected. In this short communication is presented the possible role of telomerase in pituitary adenomas. In our experience, hTERT expression correlate with cellular proliferation associated with angiogenesis and hormonal activity. In pituitary adenomas, telomerase detection must be correlating with histopathological, ultrastructural and clinical evaluation, expression of proliferation markers, cell cycle molecules, and over all with hormonal activity. The evaluation of these variables with the modern technical tools in molecular biology, as proteomic analysis, could provide information about their biological behavior.

Telomeres are DNA-protein structures located at the end of chromosomes in eukaryotic cells [1]. Human chromosomes telomeres terminate with several kilobases of the simple telomere hexanucleotide repeat d(TTAGGG)n. Recently, based on the click chemistry approach using a combination of NMR, CD and MALDI-TOF-MS experiments, it has been demonstrate that telomere DNA is transcribed into telomeric repeat-containing RNA (TERRA) in mammalian cells that can form a DNA-RNA hybrid type G-quadruplex structure [2,3]. These results serve for developing new anti-cancer reagents [4].

Telomeres have several roles: regulate gene expression, participate in the chromosomes replication, protect the ends of the DNA double helix from degradation by nucleases, and from unwanted DNA repair activities [5]. Telomeres can regulate the lifespan of cells. In somatic cells, telomeres shorten progressively with each successive cell division, because of the combination of nucleolytic processing. This progressive shortening is an important mechanism in the timing of human cellular aging. When telomeres become sufficiently short cells undergo a growth arrest called senescence phase, then p53 is upregulated, and is followed by apoptosis and death [6].

Telomerase is a specialized ribonucleoprotein polymerase that contains RNA molecule (hTR), which is used as a template segment that directs the synthesis of telomeric repeats at the 3' ends of chromosomes by the addition of simple TG- rich repeats [7,8]. The vast majority of normal somatic cells either do not express telomerase activity or express it at very low levels. In contrast, telomerase is expressed and functionally active in 80-90% of human cancers [9]. It is expressed in neoplasias as colorectal cancer, [10,11] bladder, breast, and prostate [12,13]. Thus, cancer cell DNA is continuously extended or maintained by telomerase to compensate for the loss of telomeric repeats [14], and as a result the cells become immortalized [12].

In Brain telomerase is expressed in the vast majority of primary tumors [15]. Le (1998) reported telomerase activity in 89% of glioblastomas and 45% of anaplastic astrocytomas, but it was absent in normal brain tissues [16]. In olfactory Neuroblastomas, Meduloblastoma, Paraganglioma, and Oligodendroglioma it has been suggested that high hTERT immunoexpression can be considered a potential indicator of aggressive behavior. [17,18]. In glioblastoma tissues hTERT expression represents a simple and reliable biological diagnostic tool. [19,20]. In glial tumors telomerase may represent an indicator of progression and poor prognosis. Boldrini (2006) evaluated telomerase activity and hTERT mRNA levels by telomeric repeat amplification protocol (TRAP) assay, and reverse transcription-PCR analysis respectively, and they report a significant association between telomerase activity and hTERT mRNA expression [21]. For Ependymomas telomerase is considered as strongest predictor of outcome, and important prognostic marker [22,23]. Therefore, it has been suggest that telomerase is a potentially important biomarker and prognostic indicator in brain tumors [24], and the inhibition of telomerase provides a therapeutic strategy for the treatment of cancer [25-31].

Although the amplification of hTERT has been detected in several tumors, there is a group of neoplasias in which telomerase activity was not always been found within the specimens [32]. These tumors are thought to maintain their telomeres by a mechanism termed alternative lengthening of telomeres (ALT), and can be either telomerase positive or telomerase negative. Thus telomere length, by itself, is not a definitive indication of telomerase activity [33]. ALT is defined by the lack of detectable telomerase activity plus the capacity of limitless divisions. ALT cells usually exhibit a remarkable heterogeneous telomere length in a given cell, often ranging from >20 kb to <2 kb. ALT cells also contain ALT-associated promyelocytic leukemia bodies (ABS). In about 5% of interphase nuclei, ABS in ALT cells have been shown to contain telomeric DNA, telomere-binding proteins, as well as proteins involved in DNA recombination and replication in conjunction with promyelocytic leukemia protein. The mechanism by which ALT cells maintain their telomeres is based on homologous recombination between imperfectly matched sequences [34], which is inhibited by mismatch repair proteins (MMR) [35], and so MMR defects increase homologous recombination between diverged sequences [36]. In human colon cancer cells it has been shown that telomere elongation without the reappearance of telomerase activity can occur in a cancer cell line with a MMR defect when telomerase is inhibited [37]. This ALT-like elongation event is associated with the presence of an ongoing homologous recombination process between sister chromatids (telomeric sister chromatid exchange [T-SCE]), and T-SCE is a characteristic of ATL cells. Furthermore, cancer cells responding with T-SCE after telomerase inhibition are less tumorigenic [38]. In glioblastoma multiforme tumors Yu-Jen (2006) correlated TP53 status of gliomas with telomere maintenance and patient outcome. They show that mutant TP53 correlates strongly with the ALT mechanism and good prognosis, whereas, for patients with telomerase-positive tumors, mutant TP53, confers a worse prognosis [39].

Telomerase in Pituitary Adenomas

The pituitary adenomas are relatively common tumors; they represent 10-15% of intracranial neoplasias [40]. There is no apparent relationship with gender and they can occur in any group of age. Clinically, they can cause local symptoms such as visual alterations, cephalea, endocraneal hypertension, cranial nerves affection, because of the mass effect on the neighboring structures, and hormone levels alterations showing endocrine symptoms as amenorrhea, galactorrhea, gigantism or acromegaly, Cushing's disease, and Nelson's syndrome. So the symptomatology depends mainly on its size, and on the capability of the adenoma to produce hormones [41].

Histologically, they are considered "non-invasive benign" tumors that do not show cellular pleomorfism or mitosis figures. However, in some cases they show rapid growth and can spread to nearby structures, have recurrence, be clinically aggressive and even malignize [42]. Low proportion of pituitary adenomas (< 0.5%) is classified as carcinomas [43]. These tumors have higher mitotic indices than benign adenomas, and are diagnosed by the presence of craniospinal and/or systemic metastases [44].

Pituitary adenoma origin is a complex process because of the different cellular types (6 types of cells accord its hormonal secretion), and diverse physiology of the pituitary gland. In spite of the clinical - pathological analysis and the evaluations of the expression of different molecules, aggressiveness and biological behavior of these "malignant" adenomas has not been found. It has been analyzed hormone production, cell cycle and proliferation markers; grow factors, cytokines; however until now such markers have failed to predict the aggressive behavior in pituitary adenomas [45-49].

It has been describe the enzymatic subunit of telomerase (hTERT) as an important prognostic marker. The expression of hTERT in pituitary tumors has been few analyzed. There are reports in which telomerase activity has not been detected in pituitary adenomas. Hiraga et. al (1998) analyzed telomerase activity in brain tumors. They report telomerase activity in grade II astrocytomas, anaplastic astrocytomas, glioblastomas, anaplastic oligoastrocytomas, neuroblastomas, and oligodendrogliomas. The telomerase activity in these tumors correlates with potential tumor progression and malignancy. But in contrast they did not found telomerase activity in benign tumors as meningiomas, hemangioblastomas, shwannomas, and pituitary adenomas [50].

Yoshino et. al. (2003) assessed the telomerase activity in 31 pituitary adenomas samples. They evaluated the expression by polymerase chain reaction (PCR)-based telomeric repeat amplification protocol (TRAP) assay. Telomerase expression was detected in 13% of the adenomas (4 tumor tissues, 3 patients). Two of them were primary adenomas and one recurrent adenoma (after 10 years of the first surgical resection). The telomerase activity levels were low levels with out statistical significance, but it is important to note that these adenomas with positive telomerase activity were large, recurrent, invasive, and functioning adenomas (prolactinoma, corticotoph, and growth hormone) [51].

In a study done by our work group [52] it was assessed the expression of the human telomerase reverse transcriptase (hTERT) catalytic fraction, by immunohistochemistry technique. Telomerase expression was detected in 28.6% adenomas (14 out of 49 tissue samples). The significant associated variables with telomerase expression were producing adenomas with prolactin, growth hormone, and ACTH positive in tissue; and high serum GH hormone levels. These results are in accord with Yoshino report [51].

Harada reports the case of a 16-year-old male with a prolactin-producing pituitary adenoma, who underwent three partial resections over a 2-year period. The tumor was initially negative to telomerase but 10 years later was submitted to a fifth surgery, and pituitary carcinoma was diagnosed. This time the tumor became telomerase positive [53].

Telomerase activity has been correlated with cellular proliferation markers. Yoshino (2003) evaluated proliferative cell index (PCI) with the MIB-1 antigen. He founded higher PCI in cases with telomerase expression than in those with out it [51]. In the same way, in our analysis (Ortiz-Plata 2007) hTERT expression correlate with cellular proliferation showing with high index of proliferating cell nuclear antigen (PCNAi) [52]. The significant variables associated with PCNA were acromegaly, positive prolactin, growth hormone, and ACTH, in tissue. In the case reported by Harada (2000) it was observed that the MIB-1 index and hTERT increased according to the histological malignancy [53]. The positive association between cellular proliferation and telomerase expression also has been observed in intracranial tumors [54-56].

On the other hand, the growth of the tumor depends on the development of new vessels (angiogenesis) among other factors. Tumor angiogenesis is a complex process that involves a series of interactions between tumor cells and endothelial cells (ECs) as was shown by Falchetti et.al. (2008) who demonstrated that telomerase upregulation by the ECs is a key requisite for GBM tumor angiogenesis [57]. In pituitary adenomas we founded the association between hTERT expression and vascular density quantified with CD34 antibody. This correlation was observed, in general, with producing adenomas [52].

So as it can be seen, little is yet known about telomerase expression and its possible role in pituitary adenoma since there are few investigation reports.

Nowadays with the modern experimental technology, there is an important advance in the knowledge of telomerase structure, function, related molecules, and involvement in cancer [58-65]. These new knowledge could be used to investigate the telomerase presence in pituitary adenomas.

Several techniques for assessing telomerase activity has been developed as telomeric repeat amplification protocol (TRAP) and its modified versions [12], and the real-time quantitative TRAP (RTQ-TRAP) that is considered the most promising [66,67]. There is other new technique recently developed in which a telomeric repeats-specific molecular beacon is used. It consists of an oligonucleotide hairpin with fluorophores and quenchers attached at its 5'- and 3'- termini in which the fluorescence signal is quenched by dissipating the energy as heat. This method permits the rapid and reliable quantification of telomerase activity [68].

In pituitary adenomas analysis, is necessary to carry out several techniques to correlate the different aspects that characterize this diverse type of tumors as histopathological, ultrastructural and clinical symptoms, expression of proliferation markers, cell cycle molecules, transcription factors, and over all associate these dates with the particular hormonal activity in each case.

At the present time, the proteomic technology has been applied to find out the differential protein expression in systems under diverse biological conditions. Also proteomics helps to discover potential tumors markers with improved sensitivities and specificities for the diagnosis, prognosis and treatment [69]. In pituitary gland, the proteoma analysis is at the beginning. With this technology it will be possible to identify the proteomic map of the pituitary gland and compare it with the different types of pituitary adenomas. The identified proteins could contribute to understand the pituitary adenomas [70-75].

There are a lot of things to do in pituitary adenomas study. Telomerase could be a marker of cellular proliferation associated with angiogenesis and hormonal activity. Evaluation of these variables as a whole with the telomerase associated proteins, applying modern technical tools in molecular biology may perhaps provide information about the biological behavior of pituitary adenomas and the telomerase role in these tumors.

References

[1] Blackburn, EH. Structure and function of telomeres. *Nature*, 1991, 350(6319), 569-73.
[2] Azzalin, CM; Reichenbach, P; Khoriauli, L; Giulotto, E; Lingner J. Telomeric repeat containing RNA and RNA surveillance factors at mammalian chromosome ends. *Science*, 2007, 318, 798-801.
[3] Schoeftner, S; Blasco, MA. Developmentally regulated transcription of mammalian telomeres by DNA-dependent RNA polymerase II. *Nat Cell Biol.*, 2008, 10, 228-236.
[4] Xu, Y; Suzuki, Y; Kaminaga, K; Komiyama, M. Molecular basis of human telomere DNA/RNA structure and its potential application. *Nucleic Acids Symposium*, 2009, 53, 63-64.
[5] Palm, W; de Lange, T. How shelterin protects mammalian telomeres. *Annu Rev Genet*, 2008, 42, 301-334.
[6] Valls, BC; Piñol, FC; Reñe, EJM; Buenestado, GJ; Viñas, SJ. Telomerase activity and telomere length in the colorectal polyp-

carcinoma sequence. *REV ESP ENFERM DIG (Madrid)*, 2009, 101(3), 179-186.
[7] Greider, CW; Blackburn, EH. Identification of a specific telomere terminal transferase activity in Tetrahymena extracts. *Cell*, 1985, 43(2 Pt 1), 405-13.
[8] Xu, Y; Suzuki, Y; Kaminaga, K; Komiyama, M. Molecular basis of human telomere DNA/RNA structure and its potential application. *Nucleic Acids Symposium Series*, 2009, 53, 63-64.
[9] Bechter, OE; Zou, Y; Wright, WE; Shay, JW. Telomeric recombination in mis atch repair deficient human cancer cells after telomerase inhibition. *Cancer Res*, 2004, 15, 3444-3451.
[10] Bautista, CV; Felis, CP; Espinet, JM; Salas, J. Telomerase activity is a prognostic factor for recurrence and survival in rectal cancer. *Dis Colon Rectum*, 2007, 50(5), 611-20.
[11] Saleh, S; Lam, AK; Ho, YH. Real-time PCR quantification of human telomerase reverse transcriptase (hTERT) in colorectal cancer. *Pathology*, 2008, 40, 25-30.
[12] Kim, NW; Piatyszek, MA; Prowse, KR; Harley, CB; West, MD; Ho, PL; Coviello, GM; Wright, WE; Weinrich, SL; Shay, JW. Specific association of human telomerase activity with immortal cells and cancer. *Science*, 1994, 266(5193), 2011-5.
[13] Bièche, I; Nouguès, C; Paradis, V; Olivi, M; Bedossa, P; Lidereau, R; Vidaud, M. Quantification of hTERT gene expression in sporadic breast tumors with a real-time reverse transcription-polymerase chain reaction assay. *Clin cancer Res.*, 2000, 6, 452-459.
[14] Counter, CM; Avilion, AA; LeFeuvre, CE; Stewart, NG; Greider, CW; Harley, CB; Bacchetti, S. Telomere shortening associated with chromosome instability is arrested in immortal cells which express telomerase activity. *EMBO J*, 1992, 11, 1921-1929.
[15] Rahman, R; Heath, R; Grundy, R. Cellular immortality in brain tumours: an integration of the cancer stem cell paradigm. *Biochim Biophys Acta*, 2009, 1792(4), 280-8.
[16] Le, S; Zhu, JJ; Anthony, DC; Greider, CW; Black, PM. Telomerase activity in human gliomas. *Neurosurgery*, 1998, 42, 1120-1125.
[17] Wang, SL; Chen, WT; Li, SH; Li, SW; Yang, SF; Chai, CY. Expression of human telomerase reverse transcriptase and cyclin-D1 in olfactory neuroblastoma. *APMIS*, 2007, 115(1), 17-21.

[18] Isa, MN; Sulong, S; Sidek, MR; George, PJ; Abdullah, JM. Telomerase activity in Malaysian patients with central nervous system tumors. *Southeast Asian J Trop Med Public Health*. 2003, 34(4), 872-6.
[19] Shervington, A; Patel, R; Lu, C; Cruickshanks, N; Lea, R; Roberts, G; Dawson, T; Shervington, L. Telomerase subunits expression variation between biopsy samples and cell lines derived from malignant glioma. *Brain Res.*, 2007, 1134(1), 45-52.
[20] Nakatani, K; Yoshimi, N; Mori, H; Yoshimura, Ahin-ichi; Sakai, H; Shinoda, J; Sakai, N. The significant role of telomerase activity in human brain tumors. *Cancer*, 1997, 80, 471-6,
[21] Boldrini, L; Pistolesi, S; Gisfredi, S; Ursino, S; Ali, G; Pieracci, N; Basolo, F; Parenti, G; Fontanini, G. Telomerase activity and hTERT mRNA expression in glial tumors. *Int J Oncol*, 2006, 28(6), 1555-60.
[22] Tabori, U; Ma, J; Carter, M; Zielenska, M; Rutka, J; Bouffet, E; Bartels, U; Malkin, D; Hawkins, C. Human telomere reverse transcriptase expression predicts progression and survival in pediatric intracranial ependymoma. *J Clin Oncol*, 2006, 24(10), 1522-8.
[23] Tabori, U; Wong, V; Ma, J; Shago, M; Alon, N; Rutka, J; Bouffet, E; Bartels, U; Malkin, D; Hawkins, C. Telomere maintenance and dysfunction predict recurrence in paediatric ependymoma. *Br J Cancer*, 2008, 99(7), 1129-35.
[24] Kim, CH; Cheong, JH; Bak, KH; Kim, JM; Oh, SJ. Prognostic implication of telomerase activity in patients with brain tumors. *J Korean Med Sci*, 2006, 21(1), 126-30.
[25] Marian, CO; Cho, SK; McEllin, BM; Mahe,r EA; Hatanpaa, KJ; Madden, CJ; Mickey, BE; Wright, WE; Shay, JW; Bachoo, RM. The telomerase antagonist, imetelstat, efficiently targets glioblastoma tumor-initiating cells leading to decreased proliferation and tumor growth. *Clin Cancer Res.*, 2010, 16(1), 154-63.
[26] Rossi, A; Russo, G; Puca, A; La Montagna, R; Caputo, M; Mattioli, E; Lopez, M; Giordano, A; Pentimalli, F. The antiretroviral nucleoside analogue Abacavir reduces cell growth and promotes differentiation of human medulloblastoma cells. *Int J Cancer*, 2009, 125(1), 235-43.
[27] Wong, VC; Ma, J; Hawkins, CE. Telomerase inhibition induces acute ATM-dependent growth arrest in human astrocytomas. *Cancer Lett*, 2009, 274(1), 151-9.
[28] Patel, R; Shervington, L; Lea, R; Shervington, A. Epigenetic silencing of telomerase and a non-alkylating agent as a novel therapeutic approach for glioma. *Brain Res.*, 2008, 10, 1188, 173-81.

[29] Khaw, AK; Silasudjana, M; Banerjee, B; Suzuki, M; Baskar, R; Hande, MP. Inhibition of telomerase activity and human telomerase reverse transcriptase gene expression by histone deacetylase inhibitor in human brain cancer cells. *Mutat Res.*, 2007, 625(1-2), 134-44.

[30] You, Y; Pu, P; Huang, Q; Xia, Z; Wang, C; Wang, G; Yu, C; Yu, JJ; Reed, E; Li, QQ. Antisense telomerase RNA inhibits the growth of human glioma cells in vitro and in vivo. *Int J Oncol*, 2006, 28(5), 1225-32.

[31] Ozawa, T; Gryaznov, Sergei M; Hu, LJ; Pongracz, K; Santos. RA: Bollen, AW; Lamborn, KR; Deen, DF. Antitumor effects of specific telomerase inhibitor GRN163 in human glioblastoma xenografts. *Neuro-Oncology*, 2004, 218-226.

[32] Shervington, A; Patel, R. Differential hTERT mRNA processing between young and older glioma patients. *FEBS Lett*, 2008, 582(12), 1707-10.

[33] Hakin-Smith, V; Jellinek, DA; Levy, D; Carroll, T; Teo, M; Timperley, WR; McKay, MJ; Reddel, RR; Royds, JA. Alternative lengthening of telomeres and survival in patients with glioblastoma multiforme. *Lancet*, 2003, 361, 836-838.

[34] Dunham, MA; Neumann, AA; Fasching, CL; Reddel, RR. Telomere maintenance by recombination in human cells. *Nat Genet*, 2000, 26, 447-50.

[35] Rayssiguier, C; Thaler, DS; Radman, M. The barrier to recombination between Escherichia coli and Salmonella typhimurium is disrupted in mismatch-repair mutants. *Nature (Lond)*, 1989, 342, 396-401.

[36] Elliott, B; Jasin, M. Repair of double-strand breaks by homologous recombination in mismatch repair-defective mammalian cells. *Mol Cell Biol.*, 2001, 21, 2671-82.

[37] Bechter, OE; Zou, Y; Wright, WE; Shay, JW. Telomeric recombination in mismatch repair deficient human colon cancer cells after telomerase inhibition. *Cancer Res.*, 2004, 64(10), 3444-3451.

[38] Stewart; SA; Hahn, WC; O'Connor, BF; Banner, EN; Lundberg, AS; Modha, P; Mizuno, H; Brooks, MW; Fleming, M; Zimonjic, DB; Popescu, NC; Weinberg, RA. Telomerase contributes to tumorigenesis by a telomere length-independent mechanism. *Proc Natl Acad Sci USA*, 2002, 99, 12606-11.

[39] Chen, YJ; Hakin-Smith, V; Teo, M; Xinarianos, GE; Jellinek, DA; Carroll, T; McDowell, D; MacFarlane, MR; Boet, R; Baguley, BC; Braithwaite, AW; Reddel, RR; Royds, JA. Association of Mutant TP53

with Alternative Lengthening of Telomeres and Favorable Prognosis in Glioma. *Cancer Res.*, 2006, 66(13), 6473-6476.
[40] Hall, WA; Luciano, MG; Doppman, JL; Patronas, NJ; Oldfield, EH. Pituitary magnetic resonance imaging in normal human volunteers: occult adenomas in the general population. *Ann Intern Med*, 1994, 120(10), 817-820.
[41] Kovacs, K; Horvath, E; Vidal, S. Classification of pituitary adenomas. *J Neurooncol*, 2001, 54, 121-127.
[42] Scheithauer; BW; Kovacs, KT; Laws, ER; Randall, RV. Pathology of invasive pituitary tumors with special reference to functional classification. *J Neurosurg*, 1986, 65(6), 733-744.
[43] Kaltsas, GA; Nomikos, P; Kontogeorgos; Buchfelder, GM; Grossman, AB. Clinical review: Diagnosis and management of pituitary carcinomas. *J Clin Endocrinol Metab*, 2005, 90(5), 3089-99.
[44] Doniach, I. Pituitary carcinoma. *Clin Endocrinol (Oxf)*, 1992, 37(2), 194-195.
[45] Haedo, MR; Gerez, J; Fuentes, M; Giacomini, D; Páez-Pereda, M; Labeur, M; Renner, U; Stalla, GK; Arzt, E. Regulation of pituitary Function by cytokines. *Horm Res.*, 2009, 72(5), 266-74.
[46] Tanase, CP; Neagu, M; Albulescu, R. Key signaling molecules in pituitary tumors. *Expert Rev Mol Diagn*, 2009, 9(8), 859-77.
[47] Pertuit, M; Barlier, A; Enjalbert, A; Gérer,d, C. Signalling pathway alterations in pituitary adenomas: involvement of Gsalpha, cAMP and mitogen-activated protein kinases. *J Neuroendocrinol*, 2009, 21(11), 869-77.
[48] Salehi, F; Agur, A; Scheithauer, BW; Kovac, K; Loyd, RV; Cusimano M. Ki-67 in pituitary neoplasms: a review-part I. *Neurosurgery*, 2009, 65(3), 429-37.
[49] Cakir, M; Grossman, AB. Targeting MAPK(Ras/ERK) and PI3K/Akt pathways in pituitary tumorigenesis. *Expert Opin Ther Targets*, 2009, 13(9), 1121-34.
[50] Hiraga, S; Ohnishi, T; Izumoto, S; Miyahara, E; Kanemura, Y; Matsumura, H; Norio, A. Telomerase Activity and Alterations in Telomere Length in Human Brain Tumors. *Cancer Res.*, 1998, 58, 2117-2125.
[51] Yoshino, A; Katayama, Y; Fukushima, T; Watanabe, T; Komine, C; Yokoyama, T; Kusama, K; Moro, I. Telomerase activity in pituitary adenomas: significance of telomerase expression in predicting pituitary adenoma recurrence. *J Neurooncol*, 2003, 63, 155-62.

[52] Ortiz-Plata, A; Tena-Suck, ML; López-Gómez, M; Heras, A: Sánchez-García, A. Study of the telomerase hTERT fraction, PCNA and CD34 expression on pituitary adenomas. Association with clinical and demographic characteristics. *J Neurooncol*, 2007, 84(2), 159-66.

[53] Harada, K; Arita, K; Kurisu, K; Tahara, H. Telomerase activity and the expression of telomerase components in pituitary adenoma with malignant transformation. *Surg Neurol*, 2000, 53, 267-74.

[54] Cabuy, E; Ridde,r, L. Telomerase activity and expression of telomerase reverse trancriptase correlated with cell proliferation in meningiomas and malignant brain tumors in vivo. *Virchows Arch*, 200, 1439(2), 176-84.

[55] Tabori, U; Wong, V; Ma, J; Shago, M; Alon, N; Rutka, J; Bouffet, E; Bartels, U; Malkin, D; Hawkins, C. Telomere maintenance and dysfunction predict recurrence in paediatric ependymoma. *Br J Cancer*, 2008, 99(7), 1129-35.

[56] Maes, L; Kalala, JP; Cornelissen, M; De Ridder, L. PCNA, Ki-67 and hTERT in residual benign meningiomas. *In Vivo*, 2006, 20(2), 271-5.

[57] Falchetti, ML; Mongiardi, MP; Fiorenzo, P; Petrucci, G; Pierconti, F; D'Agnano, I; D'Alessandris, G; Alessandri, G; Gelati, M; Ricci-Vitiani, L; Maira, G; Larocca, LM; Levi, A; Pallini, R. *Int J Cancer*, 2008, 122(6), 1236-42.

[58] Matsumura, Y; Shimada, K; Tanaka N; Fujimoto, K; Konishi, N. Phosphorylation status of fas-associated death domain-containing protein regulates telomerase activity and strongly correlates with prostate cancer outcomes. *Pahtobiology*, 2009, 76(6), 293-302.

[59] Rhee, DB; Wang, Y; Mizesko, M; Zhou, F; Haneline, L; Liu, Y. FANCC suppresses short telomere-initiated telomere sister chromatid exchange. *Hum Mol Genet*, 2010, [Epub ahead of print].

[60] Onitake, Y; Hiyama, E; Kamei; N; Yamaoka, H; Sueda, T; Hiyama, K. Telomere biology in neuroblastoma: telomere binding proteins and alternative strengthening of telomeres. *J Pediatr Surg.*, 2000, 44(12), 2258-66.

[61] Bhattacharyya, S; Sandy, A; Groden, J. Unwinding proteins complexes in ALTernative telomere maintenance. *J Cell Biochem*, 2010, 109(1), 7-15.

[62] Caslini, C. Transcriptional regulation of telomeric non-coding RNA: Implications on telomere biology, replicative senescence and cancer. *RNA Biol.*, 2010, Jan 6, 7(1), (Epub ahead of print).

[63] Clarke, CJ; Hill, LL; Bolden, JE; Johnstone, RW. Inducible activation of IFI 16 results in suppression of telomerase activity, growth suppression snd induction of cellular senescence. *J Cell Biochem*, 2010, 109(1), 103-12.

[64] Chebel, A; Rouault, JP; Urbanowicz, I; Baseggio L; Chien, WW; Salles, G; Ffrench, M. Transcriptional activation of hTERT, the human telomerase reverse transcriptase, by nuclear factor of activated T cells. *J Biol Chem.*, 2009, 284(51), 35725-34.

[65] Chen, YJ; Campbell, HG; Wiles, AK; Eccles, MR; Reddel, RR; Braithwaite, AW; Royds, JA. PAX8 regulates telomerase reverse transcriptase and telomerase RNA component in glioma. *Cancer Res.*, 2008, 68(14), 5724-32.

[66] Jakupciak, JP. Real-time telomerase activity measurements for detection of cancer. *Expert Rev Mol Diagn*, 2005, 5, 745-753.

[67] Murphy, J; Bustin, SA. Reliability of real-time-reverse-transcription PCR in clinical diagnostics: gold standard or substandard? *Exp Rev Mol Diagn*, 2009, 9(2), 187-197.

[68] Kong, D; Jin, Y; Yin, Y; Mi, H; Shen, H. Real-time PCR detection of telomerase activity using specific molecular beacon probes. *Anal Bioanal Chem.*, 2007, 388(3), 699-709.

[69] Wong, SC; Chan, CM; Ma, BB; Lam, MY; Choi, GC; Au, TC; Chan, AS; Chan, AT. Advanced proteomic technologies for cancer biomarker discovery. *Expert Rev Proteomics*, 2009, 6(2), 123-34.

[70] Beranova-Giorgianni, S; Giorgianni, F; Desiderio, DM. Analysis of the proteome in the human pituitary. *Proteomics*, 2002, 2(5), 534-42.

[71] Zhan, X; Desiderio, DM. Heterogeneity analysis of the human pituitary proteome. *Cell Mol Biol (Noisy-le-grand)*, 2003, 49(5), 689-712.

[72] Zhan, X; Giorgianni, F; Desiderio, DM. Proteomics analysis of growth hormone isoforms in the human pituitary. *Proteomics*, 2005, 5(5), 1228-41.

[73] Evans, CO; Moreno, CS; Zhan, X; McCabe, MT; Vertino, PM; Desiderio, DM; Oyesiku, NM. Molecular pathogenesis of human prolactinomas identified by gene expression profiling, RT-qPCR, and proteomic analyses. *Pituitary*, 2008, 11(3), 231-45.

[74] Moreno, CS; Evans, CO; Zhan, X; Okor, M; Desiderio, DM; Oyesiku, NM. Novel molecular signaling and classification of human clinically nonfunctional pituitary adenomas identified by gene expression profiling and proteomic analyses. *Cancer Res*, 2005, 65(22), 10214-22.

[75] Zhan, X; Desiderio, DM. A reference map of a human pituitary adenoma proteome. *Proteomics*, 2003, 3(5), 699-713.

In: Telomerase: Composition, Functions ...　　ISBN: 978-1-61668-957-5
Editor: Aiden N. Gagnon, pp. 95-103　　© 2010 Nova Science Publishers, Inc.

Chapter V

Regulations of Telomerase Activity and *WRN* Gene Expression

Fumiaki Uchiumi[1*], *Yoshikazu Higami*[2]
and Sei-ichi Tanuma[3,4]
[1] Department of Gene Regulation,
[2] Department of Molecular Pathology and Metabolic Disease,
[3] Department of Biochemistry & Molecular Biology, Faculty of Pharmaceutical Sciences,
[4] Genome and Drug Research Center, Tokyo University of Science, 2641 Yamazaki, Noda, Chiba 278-8510, Japan

Abstract

Telomeres, which are the specific structures at the ends of chromosomes, play an important role in regulating genome stability and cellular senescence. Telomerase is a telomere-elongating enzyme composed of TERT and TERC (TR) that are protein and RNA subunits, respectively. Besides telomerase, factors involved in DNA repair synthesis factors have also been shown to regulate telomere length. Among the DNA synthesis factors, WRN, which belongs to the RecQ

---

[*] Corresponding author: Phone: E-mail address: uchiumi@rs.noda.tus.ac.jp (F. Uchiumi) +81-4-7121-3616 Fax: +81-4-7121-3608

helicase family, is implicated in telomere regulation. The mutation of the *WRN* gene causes Werner syndrome, in which, patients show genome instability accompanied with premature aging. These observations imply that senescence is controlled by telomere maintenance factors. However, caloric restriction (CR), or glucose deprivation, extends the life span of various animals. We observed the induction of telomerase activity and *WRN* gene expression in 2-deoxy-D-glucose (2DG)-treated HeLa S3 cells. In this article, we discuss the transcriptional regulation of *WRN* and *TERT* gene expressions comparing their promoter regions.

Keywords: Caloric Restriction; 2-DG; Rec Q Helicases; Resveratrol; TERT; TERC; WRN.

Introduction

Telomeres are specific structures present at chromosomal ends that are regulated and maintained by various protein factors [1]. The telomeres play an important role in protecting chromosomes from fusion and degradation [2]. Telomeric regions are not completely replicated by the DNA replication machinery composed of DNA polymerase and its auxiliary protein complex; it requires a telomerase that is a specialized cellular ribonucleoprotein reverse transcriptase composed of protein and RNA subunits TERT and TERC (TR), respectively [3,4].

Telomere shortening is associated with cellular senescence [2,5]. In addition, genetic studies of human premature syndrome suggest that proteins involved in DNA repair synthesis, construction of nuclear structures, and telomere maintenance control the aging process [6,7]. For example, Hutchinson Gilford progeria syndrome (HGPS) and Werner's syndrome (WS) are caused by mutations in *LMNA* encoding lamin A and *WRN* genes, respectively [6-8]. Telomeric regions are thought to be heterochromatic and associated with hetelochromatin protein 1 (HP1) to the nuclear matrix membrane [9]. Therefore, lamin A might also regulate positioning of telomeres at the nuclear membrane. The RecQ helicase WRN, which unfolds G-quadroruplex structures [10], has been shown to regulate telomeres [11]. These observations suggest that the aging process is probably regulated by telomere maintenance in accordance with processes that control the nuclear membrane structure and DNA damage responses.

Effect of Caloric Restriction (CR) Mimetics on Expressions of Telomerase and WRN in Hela S3 Cells

CR is an effective and reproducible treatment to elongate life spans of yeast, warms, fruit-flies, and mammals [12]. Therefore, CR might affect the expression of proteins regulating the aging-process. The compounds 2-deoxy-D-glucose (2DG) and resveratrol (Rsv) are used as CR mimetic drugs. Previously, we observed that 2DG induces expression of *TERT* and *WRN* genes by activating their promoters [13]. As a consequence, telomerase (telomere elongating) activity and the amount of WRN protein increases in 2DG-treated HeLa S3 cells. The natural polyphenolic compound Rsv increases life spans of various organisms [14]. In addition, Rsv activates SIRT1, which is an NAD^+-dependent deacetylase [15], and translocation of the forkhead box class O transcription factor FoxO3 into FB0603 cell nuclei [14]. We therefore examined the effect of Rsv on telomerase and WRN expression in HeLa S3 cells, and observed that Rsv has a positive effect on the expression of the telomerase and WRN proteins [16]. However, induction of the promoter activities of *TERT* and *WRN* genes and their protein levels after Rsv treatment were transient, reaching the highest point within 24 h. Therefore, the induction pattern of cells treated with Rsv and 2DG was different. In addition, Rsv did not affect the viability of cells, whereas 2DG diminished cell viability in a dose-dependent manner [15]. Although, the differential response of cells to CR mimetic drugs 2DG and Rsv was prominent, the transcriptional response to activation of the telomerase and the *WRN* promoter was similar.

Comparison of the Promoter Regions of the Human *TERT* and *WRN* Genes

Given that promoter activities of the *TERT* and *WRN* genes are similarly regulated by treatment of 2DG and Rsv, there should be a common *cis*-acting element in their 5'-upstream transcription control regions. As shown in Figure 1, TFSEARCH analysis showed that their promoter regions lack TATA and CCAAT boxes but have multiple GC boxes or Sp1 binding sequences (Table 1). Sp1 is a transcription factor that binds to the GC boxes with the consensus sequence 5'-(G/T)GGGCGG(G/A)(G/A)(C/T)-3' or 5'-(G/T)(G/A)GGCG(G/T)(G/A)(G/A)(C/T)-3' [17]. Various genes encoding cell proliferation-, survival-, and growth-regulating factors are targets of Sp1 [17]. Interestingly, the GC box/Sp1 elements are located in the 5'-upstream region of the *p21 (CDKN1A)* gene [13], which encodes the cyclin-dependent kinase inhibitor [18]. Besides Sp1, c-Myc/Max [19,20] and NFAT [21] have also been suggested to regulate *TERT* promoter activity. *WRN* promoter activity was reported to be significantly reduced in the cells of WS patients [22]. Further, it has been suggested that Sp1-mediated *WRN* promoter activity is modulated by Rb and p53 [23]. These observations imply that not only Sp1 but also other transcription factors are involved in the regulation of the *TERT* and *WRN* promoters.

Signal Transduction Systems Induced by CR Mimetics

The insulin signaling cascade is thought to play an important role in controlling life spans [24]. FoxO3 protein is phosphorylated through the insulin signalling cascade, and its modification inhibits the transcription activating function [25]. When mammalian cells encounter a low-glucose environment, the concentration of insulin is lowered to evoke intracellular signals, and consequently the FoxO3 target genes may be upregulated. Moreover, it was shown that the transcription factor FoxO3 is transported from the cytoplasm to the nucleus in response to the Rsv treatment [14]. The consensus recognition sequence 5'-TTGTTTAC-3' of the FoxO protein is located at the promoter region of various genes encoding cell cycle inhibitors, oxidative stress resistance proteins, and apoptosis-associated proteins [25]. Although the FoxO binding consensus sequences are not found in the 5'-upstream regions of the human *TERT* (263 bp) and *WRN* (417 bp) genes, FoxO3 may indirectly upregulate both *TERT* and *WRN* promoters by activating Sp1 transcription. The alternative important signal transduction system that regulates transcription of aging- or senescence-associated genes might come from mitochondrial reactive oxygen species (ROS) generation [26], which is thought to cause DNA damage. Taken together, CR could diminish the insulin signaling cascade accompanied with low mitochondrial ROS generation to induce expression of the aging/senescence-associated genes, especially in mammals.

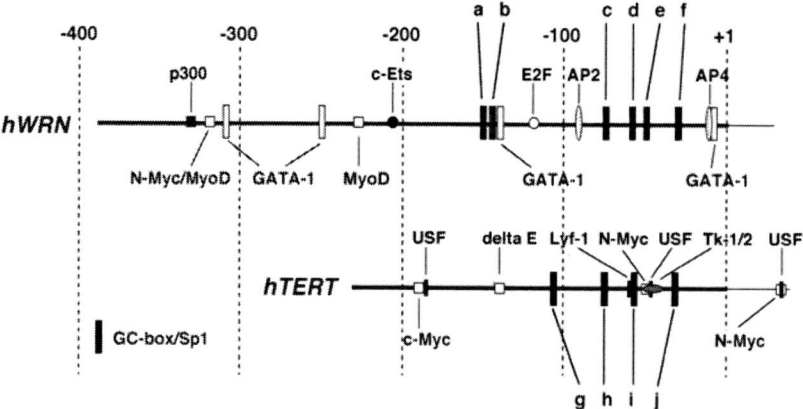

Figure 1. Comparison of the promoter regions of the human *TERT* and *WRN* genes 5'-upstream regions of the human *TERT* (263 bp) and *WRN* (417 bp) genes were isolated by PCR using genomic DNA from HeLa S3 cells as a template. After confirming the sequences, we performed TFSEARCH analysis. Symbols indicate predicted transcription factor binding sites with TFSEARCH scores over 85. Closed bars (a-j) indicate positions of GC box/Sp1 binding sequences shown in Table 1.

Table 1. Clusters of GC box/Sp1 elements in the human *WRN* and *TERT* promoter regions

Promoter	Position	sequences
WRN	a	5'-**TGGGCGGGGC**-3'
	b	5'-**GGGGCTGGAT**-3'
	c	5'-**GGGGCGGGGA**-3'
	d	5'-**GGGGCAGGAA**-3'
	e	5'-**GGGGCGGGGA**-3'
	f	5'-**GGGGCGGGTC**-3'
TERT	g	5'-**GGGGCGGGGT**-3' (inverted)
	h	5'-**GGGGCTGGGC**-3' (inverted)
	i	5'-**GGGGCTGGGA**-3' (inverted)
	j	5'-**AGGGCGGGGC**-3' (inverted)

The 10 bp consensus sequence:
5'-(G/T)**GGGCGG**(G/A)(G/A)(C/T)-3'
5'-(G/T)(G/A)**GGCG**(G/T)(G/A)(G/A)(C/T)-3'

Various cell fates can be altered experimentally by overexpression or ablation of various transcription factors [27]. It is noteworthy that the induced pluripotent stem (iPS) cells can also be generated by introducing expression vectors of four transcription factors, namely Oct4, Sox2, Klf4, and c-Myc [28]. Cellular senescence might be modulated by transcription factors that upregulate cell proliferation inhibitors and DNA/telomere repair-associated proteins. We thus propose that GC box-binding proteins and their associating factors are affected by the intracellular signals induced by CR or CR mimetic drugs.

Conclusion

CR mimetic drugs, 2DG and Rsv, up-regulate telomerase activity in HeLa S3 cells. In addition, the increase in telomerase activity is accompanied with augmentation of the *WRN* gene and protein expressions. A transient transfection experiment indicated that promoter regions of these genes positively respond to 2DG and Rsv. Comparison of the 5'-upstream regions of these two genes indicates that GC boxes or Sp1 binding sites are commonly located. Taken together, these observations imply that the GC box/Sp1 motif might be an essential *cis*-element to control gene expression of regulatory factors that are involved in the aging process.

Abbreviations

CR	Caloric Restriction
2DG	2-deoxy-D-glucose
Rsv	Resveratrol
WS	Werner Syndrome.

References

[1] de Lange, T. (2006). Mammalian Telomeres. In T., de Lange, V. Lundblad, & E. Blackburn, (Eds.), Telomeres, (second ed, 387-431). Plainview, NY: Cold Spring Harbor Laboratory Press.

[2] Blackburn, E. H. (2006). A history of telomere biology. In T., de Lange, V. Lundblad, & E. Blackburn, (Eds.), Telomeres, (second ed, 1-19). Plainview, NY: Cold Spring Harbor Laboratory Press.

[3] Morin, G. B. (1989). The human telomere terminal transferase enzyme is a ribonucleoprotein that synthesizes TTAGGG repeats. Cell, 59, 521-529.

[4] Bryan, T. M. & Cech, T. R. (1999). Telomerase and the maintenance of chromosome ends. Curr. Opin. Cell Biol., 11, 318-324.

[5] Harley, C. B., Futcher, A. B. & Greider, C. W. (1990). Telomeres shorten during aging of human fibroblasts. Nature, 345, 458-460.

[6] Oberdoerffer, P. & Sinclair, D. A. (2007). The role of nuclear architecture in genomic instability and aging. Nat. Rev. Mol. Cell Biol., 8, 692-702.

[7] Eriksson, M., Brown, W. T., Gordon, L. B., Glynn, M. W., Singer, J., Scott, L., Erdos, M. R., Robbins, C. M., Moses, T. Y., Berglund, P., Dutra, A., Pak, E., Durkin, S., Csoka, A. B., Boehnke, M., Glover, T. W. & Collins, F. S. (2003). Recurrent de novo point mutations in lamin A cause Hutchinson-Gilford progeria syndrome. Nature, 423, 293-298.

[8] Yu, C. E., Oshima, J., Fu, Y. H., Wijsman, E. M., Hisama, F., Alisch, R., Matthews, S., Nakura J., Miki, T., Ouais, S., Martin, G. M., Mulligan, J. & Schellenberg, G. D. (1996). Positional cloning of the Werner's syndrome gene. Science, 272, 258-262.

[9] Sharma, G. G., Hwang, K. K., Pandita, R. K., Gupta, A., Dhar, S., Parenteau, J., Agarwal, M., Worman, H. J., Wellinger, R. J. & Pandita, T. K. (2003). Human heterochromatin protein 1 isoforms $HP1^{Hs\alpha}$ and $HP1^{Hs\beta}$ interfere with hTERT-telomere interactions and correlate with changes in cell growth and response to ionizing radiation. Mol. Cell. Biol., 23, 8363-8376.

[10] Opresko, P. L., Mason, P. A., Podell, E. R., Lei, M., Hickson, I. D., Cech, T. R. & Bohr, V. A. (2005). POT1 stimulates RecQ helicases WRN and BLM to unwind telomeric DNA substructures. J. Biol. Chem., 280, 32069-32080.

[11] Crabbe, L., Verdun, R. E., Haggblom, C. I. & Karlseder, J. (2004). Defective telomere lagging strand synthesis in cells lacking WRN helicase activity. Science, 306, 1951-1953.

[12] Cavallini, G., Donati, A., Gori, Z. & Bergamini, E. (2008). Towards an understanding of the anti-aging mechanism of caloric restriction. Current Aging Sci., 1, 4-9.

[13] Zhou, B., Ikejima, T., Watanabe, T., Iwakoshi, K., Idei, Y., Tanuma, S. & Uchiumi, F. (2009). The effect of 2-deoxy-D-glucose on Werner syndrome RecQ helicase gene. FEBS Lett., 583, 1331-1336.
[14] Stefani, M., Markus, M. A., Lin, R. C. Y., Pinese, M., Dawes, I. W. & Morris, B. J. (2007). The effect of resveratrol on a cell model of human aging. Ann. N.Y. Acad. Sci., 1114, 407-418.
[15] Sinclair, D. A. (2005). Toward a unified theory of caloric restriction and longevity regulation. Mech. Ageing Dev., 126, 987-1002.
[16] Uchiumi, F., Watanabe, T., Hasegawa, S., Hoshi, T., Higami, Y. & Tanuma, S. Unpublished data.
[17] Wieratra, I. (2008). Sp1: Emerging roles-Beyond constitutive activation of TATA-less housekeeping genes. Biochem. Biophys. Res. Commun., 372, 1-13.
[18] Abbas, T. & Dutta, A. (2009). p21 in cancer: intricate networks and multiple activities. Nat. Rev. Cancer, 9, 400-414.
[19] Cong, Y. S., Wen, J. & Bacchetti, S. (1999). The human telomerase catalytic subunit hTERT: organization of the gene and characterization of the promoter. Hum. Mol. Genet., 8, 137-142.
[20] Kyo, S., Takakura, M., Taira, T., Kanaya, H., Itoh, M., Yutsudo, M., Ariga, H. & Inoue, M. (2000). Sp1 cooperates with c-Myc to activate transcription of the human telomerase reverse transcriptase gene (hTERT). Nucleic Acids Res., 28, 669-677.
[21] Chebel, A., Rouault, J. P., Urbanowics, I., Baseggio, L., Chien, W. W., Salles, G. & Ffrench, M. (2009). Transcriptional activation of hTERT, the catalytic subunit of telomerase, by NFAT. J. Biol. Chem., 284, 35725-35734.
[22] Wang, L., Hunt, K. E., Martin, G. M. & Oshima, J. (1998). Structure and function of the human Werner syndrome gene promoter: evidence for transcriptional modulation. Nucleic Acids Res., 26, 3480-3485.
[23] Yamabe, Y, Shiomamoto, A., Goto, M., Yokota, J., Sugawara, M. & Furuichi, Y. (1998). Sp1-mediated transcription of the Werner helicase gene is modulated by Rb and p53. Mol. Cell. Biol., 18, 6169-6200.
[24] van der Horst, A. & Burgering, B. M. T. (2007). Stressing the role of FoxO proteins in lifespan and disease. Nat. Rev. Mol. Cell Biol., 8, 440-450.
[25] Longo, V. D. (2009). Linking sirtuins, IGF-I signaling, and starvation. Exp. Gerontol., 44, 70-74.
[26] Sanz, A. & Stefanatos, R. K. A. (2008). The mitochondrial free radical theory of aging: A critical view. Current Aging Sci., 1, 10-21.

[27] Graf, T. & Enver, T. (2009). Forcing cells to change lineages. Nature, 462, 587-594.
[28] Takahashi, K. & Yamanaka, S. (2006). Induction of pluripotent stem cells from mouse embryonic and adult fibroblast cultures by defined factors. Cell, 126, 663-676.

In: Telomerase: Composition, Functions ... ISBN: 978-1-61668-957-5
Editor: Aiden N. Gagnon, pp. 105-144 © 2010 Nova Science Publishers, Inc.

Chapter VI

Telomerase and Telomere Dynamics in Cancer: Clinical Application for Cancer Diagnosis

Eiso Hiyama[*1] *and Keiko Hiyama*[2]
[1]Natural Science Center for Basic Research and Development and
[2]Department of Translational Cancer Research, RIRBM, Hiroshima University, 1-2-3 Kasumi, Minami-ku, Hiroshima, 734-8551, Hiroshima, Japan

Abstract

Telomerase, a critical enzyme responsible for cellular immortality, is usually repressed in somatic cells except for lymphocytes and self-renewal stem/progenitor cells, but is activated in approximately 80% of human cancer tissues. The human telomerase reverse transcriptase (TERT) is the catalytic component of human telomerase. In cancers in which telomerase activation occurs at the early stages of the disease, telomerase activity and TERT expression are useful markers for the detection of cancer cells. In other cancers in which telomerase becomes upregulated upon tumor progression, they are useful as prognostic indicators. On the other hand, in the ectopic expression of telomerase

---

[*] Corresponding author: Telephone No.: +81-82-255-5951; Facsimile No.: +81-82-257-5909; E-mail address: eiso@hiroshima-u.ac.jp.

which may precede cancer transformation, detection of telomerase in noncancerous lesion will become a useful marker detecting high-risk patients for cancer development. Moreover, progressive telomere shortening predominantly usually occurs during early carcinogenesis before telomerase activation. And then, the activated telomerase maintains telomere length stability in almost all cancer cells. Recently, methods for the in situ detection of the TERT mRNA and protein have been developed. These methods should facilitate the unequivocal detection of telomerase activated cells, even in tissues containing a background of normal telomerase-positive cells. Thus, telomere and telomerase is useful biomarker for detecting high-risk patients, early detection of cancer, or cancer progression. This review summarizes the data of various kinds of cancers and discusses clinical application of telomere and telomerase in human cancer.

Keywords: telomerase, human telomerase reverse transcriptase (TERT), fine needle aspiration, cytology, telomeric repeat amplification protocol (TRAP), diagnosis, prognosis, *in situ* hybridization, immunohistochemistry.

Introduction

Human telomeres are nucleoprotein complexes consisting of 8-15 kb of TTAGGG repeats along with specific binding proteins, and are located at each chromosomes end [1, 2]. These structures prevent chromosome termini from being recognized as double stranded DNA breaks and are essential to genomic stability [3]. In somatic cells, telomeres progressively shorten during each cell division due to the replication problem [4, 5]. Massive cell division leads to excessive telomere erosion, loss of telomere capping function, and eventually genetic instability and cellular senescence when telomeres become critically short [6]. Telomerase is an RNA-dependent DNA polymerase that is generally inactivated in normal human somatic cells, and is under control of the *human telomerase reverse transcriptase (TERT)* gene that encodes the catalytic component of telomerase. [7-9] Thus, telomere crisis, defined as events that occur when cells lose telomere function as a result of extended proliferation in the absence of telomerase, is a critical rate-limiting and promoting event for cancer progression [10, 11]. Telomeres can be maintained, in some human cells including germ cells, stem cells and lymphocytes and most human cancers, by telomerase activation [1].

Cancers are diseases characterized by unlimited proliferation and invasion into surrounding tissues or distant organs. Most somatic human cells lack telomerase activity, have a limited lifespan, and require the activation of telomerase for extended lifespan. Although telomerase activation is not always concomitant with carcinogenesis, its presence in 78% of more than 8,000 tumor samples analyzed (Figure 1). Among them, more than 80% of 7,000 adult cancers specimens excluding brain tumors, sarcomas, and childhood tumors exhibited telomerase activation, suggesting that the role of telomerase in cancer progression seems to be involved in telomere stabilization, preventing telomere exhaustion [3, 12-15]. By stabilizing telomeres and supporting the indefinite growth of most cancer cells, telomerase most certainly plays a crucial role in the progression and maintenance of tumors. Some reports revealed that telomerase activity is upregulated during mouse tumorigenesis in spite of the fact that mice have very long telomeres [16, 17].

Telomerase and oncogenic ras (H-rasV12) were then sequentially added to these senescence-overcoming cells by large T-antigen in rapid succession. Cells that expressed all three genes were capable of anchorage-independent growth *in vitro* and growth in nude mice [18]. These observations have suggested that telomerase may promote tumorigenesis independently of telomere length.

Telomere and Telomerase in Cancer Development

According to telomere hypothesis [19, 20], one of the primary functions of telomere is to protect linear chromosomes from damage and degradation. After extended passage, human normal cells, whose telomerase activity is repressed, enter an irreversible growth arrest called replicative senescence when the telomere in these cells shortens to critical lengths. In contrast, immortal cells such as many cell lines derived from human cancers, are capable of unlimited replicative potential, express telomerase, and maintain stable telomere lengths. Thus, stable telomere lengths by telomerase activation or alternative lengthening of telomere (ALT) mechanism might be as a molecular device of cellular immortalization. Indeed, for some cells, such as fibroblasts and endothelial cells, expression of TERT permit these cells to bypass replicative senescence and achieve immortalization [9, 21, 22]. Since carcinogenesis and immortalization are different events in principle [11, 23, 24], telomere shortening occurs when cancer cells do note get immortalization and telomerase activation or ALT occur in various steps during the multi-step process of carcinogenesis including premalignant lesion, early cancer, and progressing of cancer.

Commonly, in the premalignant lesions such as dysplasia or metaplasia, telomere length was shorter than that of normal cells due to the excessive number of the cell divisions. As described above, telomere attrition in postsenescent human cells eventually initiates crisis, which is accompanied by chromosomal fusions, providing evidence of increased genomic instability. [6] Indeed, a rare consequence of these changes in genomic structure is the activation of telomerase or ALT, which facilitates immortalization. However, this increased genomic instability caused by telomere shortening and loss of the protective function of telomeres may also drive malignant transformation under certain conditions. In mice lacking telomerase and heterozygous for p53, a situation that mimics postsenescent human cells, Artandi et al [25] noted an increased incidence of cancers, particularly epithelial malignancies. The karyotypes observed in tumors derived from these mice exhibited a high rate of nonreciprocal translocations common to human epithelial cancers but rare in murine tumors. Similar observations have also been made in mice deficient for both telomerase and adenomatous polyposis coli. [26].

These observations indicate that loss of chromosomal protection by telomere attrition may drive the formation of epithelial cancers. Indeed, telomere shortening and chromosomal abnormalities have been already discovered in the premalignant lesions in pancreas, colon, bladder, and prostate. For example, we studied the temporal sequence of telomere shortening and TERT expression during development of intrapapillary mucinous neoplasias (IPMNs) of the pancreas. Our results suggested that progressive telomere shortening represented an early and prevalent genetic abnormality acquired during IPMN carcinogenesis, and that TERT expression was up-regulated following substantial losses of telomeric DNA. Notably, transition from borderline to carcinoma *in situ* IPMNs appeared to be the critical stage at which telomere dysfunction occurred during IPMNs carcinogenesis. Progressive telomere shortening, together with telomerase activation, may eventually drive the full transformation of IPMNs [27, 28]. Similarly in inflammatory bowel diseases associated with cancers, while telomere length was gradually shortened by inflammation, high levels of telomerase activity were detected only in cancerous lesions [29].

In some instances in which immortalization by telomere stabilization is acquired before transformation, telomerase may already be ubiquitously expressed at the preneoplastic or *in situ* stage; while in other instances, the enzyme may be activated gradually with cancer progression [11, 23, 30].

This difference may affect the clinical utility of telomerase as a tumor marker, especially is crucial in dictating whether telomerase might be clinically useful for screening of high risk patients, early diagnosis, or prognostic purposes. Although most somatic human cells lack telomerase activity, some tissues contain specialized cells, including germ cells, lymphocytes, stem or its progenitor cells, or certain epithelial cells, that display weak levels of telomerase activity, which can be upregulated concomitantly with growth signals.

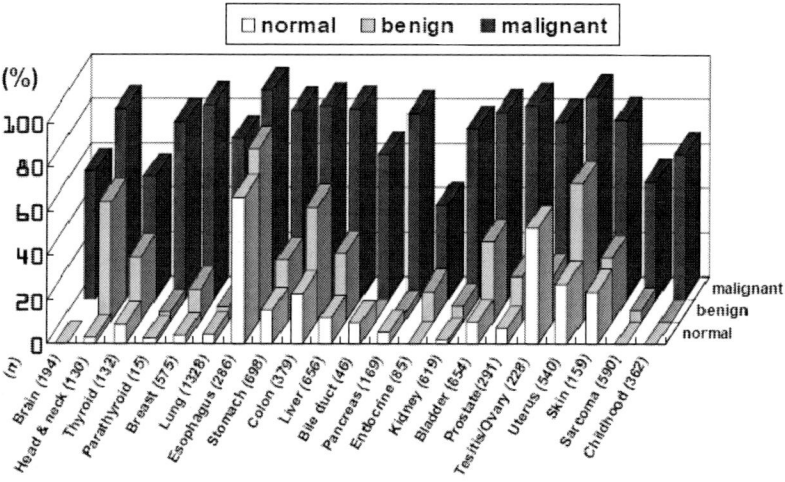

Figure 1. The percentage of telomerase or TERT positive samples in the normal, benign, malignant tumors in each organ. The percentages were calculated from the review papers and recent numerous reports [11, 12, 23, 30, 152]. Normal: normal tissues (white bars), benign tissues including premalignant regions (gray bars), and malignant cancer or sarcoma tissues (black bars):. The positive ratios of malignant tissues were more than 85% excluding brain, thyroid, bile duct, sarcoma, and childhood cancers. Interestingly, normal esophageal tissues as well as testis/ ovary showed high positive ratios. Except for these tumors, telomerase were upregualted in almost all cancer tissues, suggesting that telomerase/TERT may be a useful marker for cancer diagnosis.

In the tissues containing such cells, *in situ* detection of telomerase is needed to determine whether telomerase expression is derived from normal telomerase-positive cells or from malignant cells.

Detecting Methods of Human Telomerase/TERT in Clinical Materials of Human Cancers

Telomerase can be measured by a PCR-based assay called TRAP (Telomeric Repeat Amplification Protocol) [5], which is quite sensitive and can detect as few as 10 telomerase positive cells [31]. With this high sensitivity of these assays, telomerase activity can also be detected in certain normal somatic tissues, especially in proliferative progenitor cells and/or stem cells of self-renewing tissues (such as intestinal epithelium) and activated lymphocytes [32-34]. Moreover, the low level of telomerase activity is also detected in some benign tumors such as fibroadenomas of the breast [35], hyperplastic nodule/adenomas of the thyroid [36], and colon adenomas [33]. Generally, telomerase activity in normal somatic cells tends to be much lower in comparison to that detected in malignant cells. In clinical cancer tissues containing viable cancer cells, the levels of telomerase activity are significantly higher than those of matched control tissues but some cancer cells do not show telomerase activation [30]. Although the ratios of telomerase positive tumors are wide-ranging, the overall ratio of cancers with telomerase activity is approximately 80% (Figure 1). The survey of telomerase activity in human sarcoma revealed that telomerase activated tumors are less frequent than in cancers, partially due to the ALT (alternative lengthening of telomeres) mechanism that maintains long telomeres in some part of sarcoma. Thus, the presence of telomerase activity has been detected in the majority of cases but the frequency of tumors with detectable telomerase is variable (Figure 1).

The data obtained by the modified semi-quantitative TRAP assay, a more precise method of quantification, revealed that the levels of telomerase activity are low in normal somatic cells, however, those of activated lymphocytes showed high activity which is equivalent to that of cancer cells [32]. Accordingly, when cell samples are examined, we recommend using a thousand-cell-equivalent cell lysate per assay, to avoid false-positive results due to a contamination with lymphocytes, as proteins extracted from a thousand adult lymphocytes do not produce detectable telomerase activity [32, 37]. Moreover, to avoid false-negative results, careful attention should be paid to the degradation of telomerase and the presence of PCR inhibitors when examining clinical specimens. The survey of the malignant tumors is listed in Figure 1.

Human telomerase activity is associated with the expression of two major components: human telomerase RNA (TERC) [38] and human telomerase reverse transcriptase (TERT) [8]. Recent studies have focused on the expression of these two components as surrogates for telomerase activity and discussed their value as tumor markers. Recent studies have targeted the detection of the *TERT* mRNA or TERT expression as tumor marker for cancer diagnosis. The existence of splicing variants of the *TERT* mRNA is problematic for the use of RT-PCR for the *TERT* mRNA detection. Moreover, when using RT-PCR or the TRAP assay to detect the *TERT* mRNA or telomerase activity, the presence of normal telomerase-positive cells, such as lymphocytes or basal epithelial cells, can cause false positive results. Methodology for the *in situ* detection of telomerase in individual cells would be expected to solve this problem. An *in situ* TRAP assay was previously developed to detect the telomerase activity, but this methodology could only be used on fresh viable cells [39]. The use of *in situ* hybridization (ISH) to detect components of the telomerase complex (TERC and *TERT* mRNA), on the other hand, would be applicable to fixed tissues. However, TERC is also present at low levels in most cells lacking telomerase activity, and its level does not always correlate with telomerase activity [40]. A better target for ISH detection would be the *TERT* mRNA, whose levels appear to closely parallel to those of telomerase activity [41, 42]. *In situ* immunohistochemistry (IHC) for detection of TERT is very useful and easy for evaluation of the source of telomerase expression in a wide variety of clinical samples, including archival paraffin-embedded specimens [42-44]. Thus, we should have a good command to use these markers: telomerase activity and TERT detection.

The TERT protein can now be detected in paraffin-embedded samples and core biopsies with the use of polyclonal or monoclonal antibodies in conjunction with appropriate antigen retrieval (Figure 2) and/or the highly sensitive tyramide-based method of signal amplification [45]. Since IHC does not require specialized equipments for detection, TERT IHC is expected to become a powerful new technology for cancer detection.

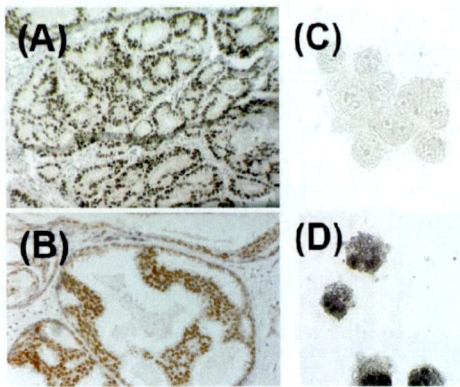

Figure 2. Immunohistochemical detection of TERT in cancer samples. An anti-TERT sera (EST21-A™, Alpha Diagnostic Int. Co., San Antonio, TX) was employed to reveal the presence of the TERT protein in a tissue sample of a clonic adenocarcinoma (A), in a tissue sample of ductal carcinoma in the breast (B), and in a brushing sample of the patient with pancreatic duct adenocarcinoma. Tumor cells are distinguished by the presence of brown pigments in the nucleus of TERT-positive cells. Staining with 3,3'-diaminobenzidine (DAB) was preformed as described previously [43, 94]. The cells obtained from brushing sample were formalin-fixed and paraffin-embedded. For both samples, heat-based antigen retrieval was performed using a citrate buffer

Telomerase/TERT for Cancer Diagnosis

Originally, acquiring immortalization by telomere stabilization with telomerase activation was seen as an independent event to carcinogenesis.

Thus, telomerase activation/TERT expression may occur in the various stages of carcinogenesis *in vivo* (Figure 3). On the other hand, there is evidence that telomerase activation may be one of the critical factors for *in vitro* transformation. Moreover, more than 78% of cancer tissues have telomerase activity, indicating that telomerase is one of the useful biomarkers in cancer. Although reported telomerase activity levels varied in human cancers, a lot of studies on the potential use of telomerase as a cancer biomarker have endorsed it as becoming a routine clinical application.

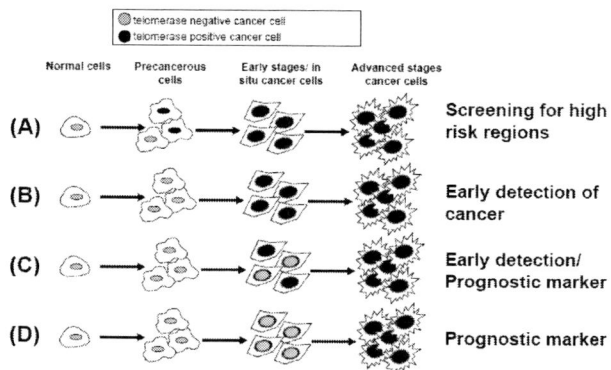

Figure 3. Scheme of telomerase activation for cancer diagnosis. Telomerase is activated in various stages of the tumor progression. (A): In the tumors whose telomerase is activated in premalignant regions, telomerase/TERT detection is useful for the screening of high risk patients. (B, C): In the tumors whose telomerase is activated concomitant with carcinogenesis, telomerase/TERT detection is useful for early detection of cancer. (C, D): When the levels of telomerase activity gradually increase with tumor progression or telomerase is activated in advanced stages of tumors, telomerase/TERT detection becomes a prognostic marker

Telomerase in the Screening for High Risk Premalignant Lesions

In the tumors whose telomerase activation occurs in pre-malignant lesions, detection of telomerase/TERT is useful as screening for high risk premalignant lesions and follow-up monitoring of such patients. Histologically "normal" bronchial epithelia in smokers may unphysiologically express telomerase as a field, and such epithelia are likely susceptible to developing lung cancer.

Ectopic expression of telomerase in bronchial epithelia may precede transformation in human lung cancer development and that detection of TERT in noncancerous bronchial epithelia will become a useful marker detecting high-risk patients for lung cancer development [46].

It indicated that [46] chronic carcinogenic insults with tobacco may result in the process of field immortalization like field cancerization, where multiple independent malignant and premalignant foci develop [47]. The detection of telomerase in premalignant or normal bronchial mucosa might be an important marker for screening of high-risk lesions. Similarly, glandular dysplasia in

Barrett's esophagus may regress spontaneously but can also progress to cancer in some cases.

Telomerase activation merits evaluation as a candidate biomarker for increased risk of persistent dysplasia and cancer progression in Barrett's esophagus [48]. Oral dysplasia revealed the same manner because the telomerase/TERT expression was detectable at the stage of precancerous oral epithelial changes [49].

Moreover, *TERT* mRNA assessment showed high sensitivity for high-grade dysplasia in the cervix [50, 51] and for high grade lesion (HGPIN) in the prostate. These data may present telomerase or TERT expression in these tissues as an appropriate target/marker for chemoprevention. [52].

Telomerase/TERT in the Early Diagnosis Cancer

In the past decade, a lot of studies were reported for the detection of telomerase activity and/or TERT expression in clinical materials as a diagnostic tool for various cancers. In a type of cancer in which telomerase is activated in early stage cancer or *in situ* carcinoma, it is a most appropriate marker for the early detection of cancer. In other types of cancer in which telomerase is activated during tumor progression, it might be a marker better suited for prognostic/malignant grade.

The many kinds of cancers showed the upregulation or reactivation of telomerase concomitant with carcinogenesis. In these kinds of cancer including breast, lung, pancreas, and prostate cancers, detection of telomerase/ TERT is useful for cancer diagnosis. Actually, the diagnostic accuracy mainly depends on the frequency of the tumors with telomerase activation (Figure 1) as well as the feasibility of the sampling of target tissues (Table 1).

Telomerase/TERT Expression in Prediction of Prognosis and Grading of Malignant Tumors

In certain types of cancers, telomerase activity is upregulated during tumor progression, so that the level of telomerase activity can be used to evaluate the malignancy grade of tumors and predict patient prognosis (Table 2). In certain cancers of the adults, the activation of telomerase correlates with advanced disease and poor prognosis, as in the cases of non-small cell lung cancer, gastric cancer, colorectal cancer, and soft tissue sarcomas [53-57]. Most representative tumors whose malignant grades correlate with telomerase activation are brain tumors and childhood malignant tumors.

Brain tumors are distinguished from other tumors because their origin is so lethal even if their histology is benign. Telomerase activity can also predict the outcomes of patients with glioma, brain tumors that are consistently difficult to be designated as either benign or malignant. Studies have shown that telomerase activity is present in most cases of malignant gliomas except for grade I gliomas, making it a useful indicator of the malignant grade of gliomas [58, 59]. In brain tumors, TERT expression was the strongest predictor of outcome and was independent of other clinical and pathologic prognostic markers [60, 61]. In malignant brain glioma and astrocytoma, there was a strong evidence of the involvement of telomerase in tumor angiogenesis [62, 63]. In pituitary adenoma, detection of telomerase expression may also correlate with biological aggressiveness and potential for regrowth [64].

Most pediatric solid tumors activate the enzyme telomerase, as shown in most adult cancers. Some pediatric tumors, including osteosarcoma and glioblastoma multiforme, lack telomerase activity and maintain telomeres via ALT (alternative lengthening of telomeres). In general, tumors showing high levels of telomerase expression are associated with an unfavorable outcome. Neuroblastoma is one of the pediatric tumors that display a well-documented relationship between tumor biology and patient outcome. In these tumors, poor prognosis is associated with high levels of telomerase activity and full-length *TERT* mRNA expression [65-70].

Interestingly, in stage 4S neuroblastoma, which represents a unique entity characterized by a high frequency of spontaneous regression, telomeres were shortened and telomerase activity was undetectable [65, 66]. In hepatoblastoma, the prognosis of the patients with high *TERT* mRNA expression or high telomerase activity was significantly worse than that of others, indicating that telomerase activation is a prognostic factor in this childhood cancer [71]. In pediatric ependimoma and soft tissue sarcoma, high telomerase is a strong predictor of outcome which was independent of other clinicopathological markers [61, 72]. These findings revealed that telomerase activation/TERT expression predicts progression and survival in most pediatric malignant tumors.

Table 1. Telomerase/ *TERT* mRNA as a diagnostic marker in clinical materials

Organs / Samples	Telomerase Activity		*TERT* mRNA	
	Cancer Positive (%)	Noncancerous Positive (%)	Cancer Positive (%)	Noncancerous Positive (%)
Excretion, Secretion, Washing or Brushing				
Oral/ Washing	110/195 (56)	70/321 (22)		
Lung/ Sputum	15/42 (36)	0/10 (0)		
Lung/ Brushing, BAL	123/188 (65)	16/211 (8)		
Colon/ Washing	20/34 (59)	0/20 (0)		
Biliary duct/ Bile	4/37 (11)	0/25 (0)	21/26 (81)	9/39 (23)
Pancreas/Pancreatic juice	59/72 (82)	2/51 (4)	10/20 (50)	0/14 (0)
Bladder/ Voiding urine	374/637 (59)	44/488 (9)	15/17 (88)	2/19 (11)
Bladder/ Washing urine	229/302 (76)	6/153 (4)	159/179 (89)	6/169 (4)
Prostate/ Voiding urine*	21/33 (64)	1/21 (5)	125/168 (74)	19/165 (12)
Uterus/ Cervical scraping	105/273 (38)	37/233 (16)	14/17 (82)	11/44 (25)
FNA				
Thyroid/ FNA				
Breast/ FNA	64/96 (67)	23/155 (15)	44/57 (77)	15/52 (29)
Mediastinal LN/ FNA	210/265 (79)	40/355 (11)	10/16 (63)	18/71 (25)
Pancreas/ FNA	18/18 (100)			
Biopsy				
Oral/ Biopsy	25/26 (96)	9/41 (22)		
Lung/ Biopsy	86/128 (67)	0/10 (0)		
Esophagus/ Biopsy	52/54 (96)	33/48 (69)		
Stomach/ Biopsy	23/29 (79)	10/28 (36)		
Colon/ Biopsy	110/126 (87)	57/148 (39)	47/58 (81)	11/13 (85)
Liver/ Biopsy	53/86 (62)	17/58 (29)	88/101 (87)	62/192 (32)
Biliary duct/ Biopsy	20/26 (77)	0/10 (0)	32/32 (100)	17/49 (35)
Bladder/ Biopsy	46/54 (85)	30/56 (54)	21/23 (91)	17/63 (27)
Prostate/ Biopsy	130/166 (78)	19/136 (14)	6/10 (60)	0/6 (0)
Uterus/ Biopsy	138/164 (84)	58/158 (37)	83/104 (80)	1/8 (13)
Skin/ Biopsy	130/159 (82)	11/109 (10)		
Pleural effusion	175/205 (85)	20/155 (13)	14/15 (93)	6/15 (40)
Peritoneal Lavage	102/141 (72)	5/117 (4)		
Blood/ Serum	59/95 (62)	0/80 (0)	4/16 (25)	0/23 (0)

These percentages were calculated from the review papers [11, 12, 23, 152] and recent numerous reports in addition to our unpublished data.

Abbreviations: BAL, bronchoalveolar lavage; FNA, fine needle aspirates.

* Voiding urine after massage.

Table 2. Telomerase activity/ *TERT* mRNA as a prognostic marker

Tumor	Correlated with prognosis	Correlated with other markers
Central nervous system malignant lymphoma; Pituitary tumor	Telomerase [64, 153] TERT [153]	
Head and neck cancer; Oral cavity and oropharynx postchemotherapeutic tumors	Telomerase [154, 155]*	
non-small cell lung cancer	Telomerase [56, 156-158] TERT [159]	TERT and Dysplasia in smokers [160]
Breast cancer	Telomerase [74]+	Telomerase and Proliferative index [161]; telomerase and Relapse-free period [114]
Thyroid Papillary carcinoma		Telomerase andExtrathyroidal extension [162]
Gastric cancer	Telomerase [54, 163-165]	
Colon cancer	Telomerase [57]	Telomerase and Advanced stages [166, 167] TERT and Advanced stages [168-170] Telomerase and Risk for metastasis [171]
Hepatic metastasis of colorectal cancer	TERT [73]	
Hepatocellular carcinoma	Telomerase [126, 172, 173]	Telomerase and Recurrence risk [174]
Pancreas Endocrine tumors	Telomerase [175]	
Renal cell carcinoma		Telomerase and Tumor grade [176] Telomerase and Advanced stage [177]
Bladder cancer (Transitional cell carcinoma)	TERT([178, 179]	Telomerase and IanTumor relapse [180]
Prostate cancer		Telomerase/TERT and Advanced stage [181]
Endometrial carcinoma		Telomerase and Recurrence risk [182]
Osteosarcoma	Telomerase [183]	Telomerase and Response to chemotherapy [55]

Table 2. (Continued)

Tumor	Correlated with prognosis	Correlated with other markers
Soft tissue sarcoma	Telomerase [184, 185]	Telomerase/TERT and Recurrence and metastasis [79]
Liposarcomas	TERT [184]	
Neuroblastoma	Telomerase [65, 66, 68] TERT [67]	Telomerase and Cytogenetic abnormalities [65, 66, 186]
Nephroblastoma		TERT andRecurrent risk [187, 188]
Hepatoblastoma	Telomerase/TERT [71]	

* Telomerase activity.
** TERT mRNA.
+ There are some controversial reports.

In patients with colorectal cancer [57], including those undergoing curative resection of liver metastases [73], both telomerase activity and TERT mRNA expression could be used as independent prognostic factors. In a retrospective study of a large number of breast cancers, the levels of telomerase activity significantly correlated with clinical outcomes and several aggressive tumor phenotypes [74].

Recently, TERT mRNA levels were reported to be higher in patients who had recurrent disease or died from breast cancer, indicating that the levels of TERT expression may be a prognostic marker [75, 76] and a monitoring marker of treatment in breast cancer. In thyroid tumors, telomerase activity may be useful to distinguish benign from malignant tumors and might provide a useful indicator of prognosis [77, 78]. Telomerase activity correlated with the grade of malignancy and more aggressive malignant histology subtypes in soft tissue sarcoma [79].

These results suggest that telomerase activation represents a simple and reliable biologic prognostic factor for intracranial tumors, childhood tumors, and some kinds of adult tumors and take decisions on the appropriate treatments such as postoperative adjuvant chemotherapy. These findings will also stimulate research on antitelomerase drugs for treatment of malignant brain tumors.

Telomerase/Tert Detection in Clinical Materials

From 1994 when the highly sensitive TRAP method was established, a lot of studies were reported for the detection of telomerase activity and/or TERT expression in clinical materials as a diagnostic tool for various cancers. In a type of cancer in which telomerase is activated in early stage cancer or *in situ* carcinoma, it is most appropriate maker for early detection of cancer. In another type of cancer in which telomerase is activated during tumor progression, it might be a marker better suited for prognostic/malignant grade. The frequency of the tumors with telomerase activation and/or TERT detection are summarized in Figure 1 and the positive ratios of telomerase activity/ TERT expression of various clinical materials are summarized in Table 1.

Cell Samples Derived from Excretion, Secretion, Washing or Brushing

Several reports demonstrated telomerase activation in a high percentage (78-95%) of squamous carcinoma derived from head, neck, and esophageal cancer patients [80-82]. Although the cells derived from oral washings are not so viable, high telomerase activity is often detected in them in patients with oral malignancy [80, 83]. In such specimens, it is difficult to avoid contamination by substances that inhibit PCR, such as necrotic tissue, leukocytes, erythrocytes, dental plaque, and bacteria, which consequently lead to false negative results in the samples containing telomerase-positive cancer cells. Thus, the accuracy for cancer diagnosis in these lesions is higher in the samples obtained by scrapping or biopsy.

In lung carcinogenesis, activation of telomerase likely represents early events in heavy smokers [84], and sputum, bronchoalveolar lavage (BAL), bronchial brushing, and bronchial washing samples were examined for telomerase activity and feasibilities for the detection of lung cancer cells were compared [85]. The sensitivity of the telomerase assay in sputum was unsatisfactory for the detection of cancer cells because sputum contains an abundance of mucus, which interferes with PCR and other enzyme reactions [85]. On the other hand, in brushing or BAL samples, telomerase activity showed a relatively high sensitivity for the detection of lung cancer cells, especially in squamous cell carcinoma as similar in head, neck, and esophageal cancers. However, BAL samples can contain activated lymphocytes, which can give false positive results in benign diseases. The clonal expansion of lymphocytes, in particular, can produce strong telomerase activity [86], which may instead reflect the aggressiveness of autoimmunity in certain benign diseases [87]. To distinguish the origin for telomerase activity, TERT detection by IHC is also useful.

Cells derived from colon luminal washing can also be subjected to the TRAP assay for cancer diagnosis. Because washing samples rarely contain basal crypt cells, the specificity of the TRAP assay for colon washing samples was remarkable but the sensitivity was found to be relatively low [88].

In patients with pancreatic ductal adenocarcinomas, the pancreatic juice contains freshly exfoliated ductal cells that carry very high levels of telomerase activity. Because of its high sensitivity and specificity, the detection of telomerase activity and/or *TERT* mRNA in pancreatic juice has become a promising new application of cancer diagnosis [37, 89-91]. Moreover, detection of telomerase activity in pancreatic juice was additionally useful for the differential diagnosis of benign and malignant intraductal papillary mucinous neoplasms (IPMN) of the pancreas, which is difficult to be distinguished preoperatively [92, 93]. In pancreas, IHC detection of TERT protein in cells derived from pancreatic juice provides a potent method for cancer diagnosis [94]. With the exception of pancreatic juice, the sensitivity of telomerase/*TERT* detection is low for excretion and secretion samples, such as bile [95, 96]. The detection of telomerase as a biomarker of bile duct cancers remains problematic [97].

Among exfoliated materials, voided urine is the easiest materials to be tested for bladder cancer screening. For the detection of bladder cancer using voided urine samples obtained from bladder cancer patients and controls, the TRAP assay showed the highest sensitivity (67%) and specificity (99%) as screening methods [98]. Since the viability of cells in voided urine samples varies, the sensitivity of telomerase activity for cancer diagnosis was lower than specificity. As an alternative, high sensitivity could be obtained in urine samples by the detection of *TERT* mRNA with RT-PCR [99, 100] or real time RT-PCR [101]. While telomerase activity and *TERT* mRNA might both be useful for the detection of bladder cancers in bladder washing samples, detection of the *TERT* mRNA may be preferable for the screening of voided urine. [99, 102]. Using the biopsy samples, both of evaluating telomerase activity and/or *TERT* gene expression levels could be used as a marker of malignant progression, useful in the early diagnosis and follow-up of bladder carcinomas [103].

In voided urine samples obtained after prostate massage, the sensitivity of telomerase activity was higher than that of cytological examination for the detection of prostate cancer [104]. As a surrogate for unstable telomerase, *TERT* mRNA was an even more reliable marker. Some invesigators reported that one of the clinical benefits resulting from the use of this new assay would be to refine the biopsy indication and to avoid for several patients without the unnecessary cost and the complications of prostate biopsy [105]. The needle biopsy (SNB) specimens from the prostate glands showed higher sensitivity and specificity [106, 107].

Fine Needle Aspirates

In thyroid and breast lesions, fine needle aspiration (FNA) is widely used as a diagnostic tool for cancer detection because these lesions are easily palpable. Moreover, for the tumors in the thyroid gland, differential diagnosis between follicular adenoma and adenocarcinoma is difficult by morphological FNA cytology alone. The detection of telomerase activity and/or *TERT* has been found to be a useful tool for this differential diagnosis since cancers gave positive signals while adenomas were negative for telomerase expression [108, 109]. However, careful attention should be paid in thyroid tissues containing lymphocytes. Telomerase activity and *TERT* expression from these inflammatory cells may also be detectable in certain benign diseases, such as Hashimoto thyroiditis [77]. Thus, detection of telomerase and *TERT* mRNA in FNA samples showed high sensitivity but low specificity [110, 111]. To distinguish the origin of telomerase, TERT detection in smear or cytospin samples is feasible and useful for cancer diagnosis. The discrimination of TERT positive cells can be useful to distinguish between benign and malignant follicular lesions of the thyroid. [112] In breast lesions, normal mammary tissue does not show detectable telomerase activity, while the activity/TERT expression is detected in 80–90% of ductal carcinoma *in situ* (DCIS) lesions and 90% of invasive breast cancers [35, 113, 114]. When we use telomerase as a biomarker for breast cancer detection, one of the most common problems is the presence of telomerase activity in benign fibroadenomas. Approximately 40% of fibroadenoma tissues display low-level telomerase activity and *TERT* mRNA expression [35, 115-117]. In combination with cytology, the screening of FNA samples for telomerase/TERT expression with careful attention to benign diseases is likely to become a powerful tool for the detection of breast cancers [43, 115, 118-

120]. Our previous results revealed that some patients whose breast suspicious lesions had shown telomerase activity in FNA samples were followed up and later diagnosed as having cancers in these lesions [11]. Thus, telomerase/TERT expression is also one of the useful diagnostic markers in breast cancer.

Biopsy Samples

In tumor biopsies from head, neck, and esophageal cancer patients, telomerase activity and the *TERT* mRNA are almost always detected at high levels in cancer cells, while low levels of telomerase activity are detectable in the normal epithelia and in approximately 20% of non-cancerous biopsy samples. Thus, a quantitative TRAP assay may be required for cancer diagnosis when scrapping or biopsy samples are examined. Moreover, some reports revealed a rather high frequency of telomerase activation in precancerous/benign tissue specimens and the adjacent normal tissue specimens [81, 82, 121]. The detection of telomerase in premalignant or normal mucosa might be an important marker for screening of high-risk lesions as described later.

In the digestive organs, biopsy by endoscopic examination is the routine diagnostic tool for the cancers. In the esophagus, high telomerase activity was detected in squamous cancer and adenocarcinoma of Barrette's esophagus [40, 122]. In the stomach, adenomas did not show telomerase activation but most gastric cancers did [123]. In intestine, telomerase activity was also detectable at the low levels in majority of normal digestive tissues, where the telomerase/TERT expression can be found in the normal basal cells of crypts [33]. Thus, a more precise measurement of the level of telomerase activity (i.e. quantitative telomerase assay) in biopsy samples of the gut may be necessary for the diagnosis of cancers.

In the liver, there is a remarkable capability to restore its functional capacity following liver injury. Therefore, in the context of liver regeneration, telomerase activation might be a cellular mechanism to confer an extended lifespan to replicating hepatocytes and hepatic progenitor cells. On the other hand, high levels of telomerase activity are a hallmark of cancer, including hepatocellular carcinoma, and hepatoblastoma, albeit low-level expression of these markers was also reported in non-cancerous tissues [71, 124]. In liver biopsy specimens, telomerase activity is useful in the differential diagnosis of hepatocellular carcinoma [125, 126]

In the pancreas, detection of telomerase activity or *TERT* mRNA in biopsy samples also displayed a high sensitivity for cancer diagnosis.

In the uterus, as a potential biomarker of cervical dysplasia, telomerase has also been the focus of intense investigations. In cervical cancers, whether telomerase is activated in pre-malignant lesions remains controversial. According to several studies published on cervical biopsies [127-130], telomerase activity is abnormally present in a remarkably high proportion of high-grade squamous intraepithelial lesions (HSILs), indicating that the activation of telomerase is an early event in the malignant progression of cervical lesions. Frost et al. have observed changes in the tissue distribution of the TERT protein in cervical cancers. While the TERT protein was limited to the lower suprabasal cells in normal cervix, it was present at all levels of the regional cells in moderate to severe dysplasia [45]. These may be the causes for the controversial results for cervical cancers. Neither detection of telomerase activity or its components, nor detection of high risk HPV seem unsuitable for the triage of women with borderline, mild, and moderate cytological dyskeratosis [131]. In radical hysterectomy samples of cervix, telomerase activity was associated with the presence of histological malignancy but there was no association between the telomerase activity and the presence of HPV [130].

Skin is a surface organ from which biopsy specimens are easily prepared. Investigations of telomerase activity as a marker of skin cancer showed that epidermal basal cells had low levels of telomerase activity; that telomerase was not activated in the vast majority of squamous cell carcinoma; but that most cutaneous malignant melanomas displayed high-levels of telomerase activity [132].Telomerase expression was reported to be correlated to critical parameters of malignant melanoma [133], although telomerase activity was weaker in Spitz nevi[134]. To elucidate the correlation between carcinogenesis and telomerase activation in the skin, further studies on skin cancers and related lesions should be done.

In soft tissue tumors, the positive ratios of detectable telomerase were approximately 20-70%, which were less than those reported in cancer [135-140]. The ALT (alternative lengthening of telomeres) pathway was reported to be more commonly activated in tumors of mesenchymal origin.[141]

Pleural Effusion and Peritoneal Lavage

Several attempts to detect telomerase activity or *TERT* mRNA in pleural effusion and peritoneal lavage samples have been reported [142-145]. Because carcinomas from almost any tumor sites can metastasize to the pleura or peritoneum, pleural effusions or ascites may contain cancer cells originating from various organs such as breast, ovary, or gastrointestinal tract. In malignant pleural effusions diagnosed by either fluid cytology or pleural biopsy, Yang et al. 1998[145] detected telomerase activity in 91% of cases with a specificity of 94%, indicating that the measurement of telomerase activity is a useful adjunct to cytology for detecting cancer cells.

Because peritoneal dissemination usually occurs in advanced stages of digestive cancers, telomerase activity in peritoneal lavage samples also showed a high specificity for cancer cell detection [146].

Both the sensitivity and specificity of the telomerase assay were higher than those of cytology for diagnosis of the malignancy, remaining only a few false positive samples such as from patients with tuberculosis or lymphocytic contamination. Duggan et al. also found telomerase activity to be more sensitive than cytology in ascitis obtained from patients with ovarian cancer [147]. Thus, in samples derived from pleural effusions and ascitis, the sensitivity of the telomerase assay for detecting cancer cells is relatively high.

Detecting Circulating Cancer Cells

Irrespective of the tumor type, the blood in patients with cancer is likely to contain circulating cancer cells that could potentially be detected using the telomerase assay [148]. The detection of these rare cancer cells in whole blood samples would be predicted to be masked by the potential presence of activated lymphocytes expressing high levels of telomerase activity [32, 86].

To detect circulating carcinoma cells using the telomerase assay, immunomagnetic separation can first be used to isolate epithelial cells from peripheral blood mononuclear cells, after which the harvested cells can be tested for telomerase activity. In one report, the harvested circulating epithelial cells showed telomerase activity in 70-80% of patients with advanced lung, colon, or breast cancers, suggesting that telomerase activity may become a useful clinical marker of circulating epithelial cancer cells [148].

One of the most routinely collected bodily fluids is blood serum, which can easily be prepared by centrifugation of whole blood. If tumor cells undergo necrosis and release their contents, some tumor-specific molecules might be present in serum that could be detected. Serum *TERT* mRNA could be detected in the serum samples derived from breast cancer patients and showed higher values in patients with HCC than those with chronic liver diseases, indicating that serum *TERT* mRNA is a novel and available marker for cancer diagnosis and follow-up [149-151].

Conclusion

The present chapter reviewed the use of human telomerase and TERT as a screening marker of high risk patients, as a cancer diagnostic marker for early detection and as a prognostic marker for predicting the outcome of individual patients as well as a marker that can distinguish malignancies from benign tumors and a marker for detecting circulating cancer cells in the blood. *In situ* hybridization and the immunohistochemical detection of TERT can now be used to identify telomerase-positive cancer cells in a background of non-cancerous cells. In the near future, convenient methods for the immunohistochmical detection of TERT are likely to become of common use in clinics for both the diagnosis of cancers and the grading of malignancies.

Abbreviations

TERT	human telomerase reverse transcriptase,
TRAP	telomeric repeat amplification protocol,
TERC	human telomerase RNA,
FNA	fine needle aspiration,
BAL	bronchoalveolar lavage,
ISH	in situ hybridization,
IHC	immunohistochemistry.

References

Blackburn, E. H. (1991). Structure and function of telomeres. *Nature (London)*, *350*, 569-573.
van Steensel, B. & de Lange, T. (1997). Control of telomere length by the human telomeric protein trf1. *Nature*, *385*, 740-743.
Artandi, S. E. & DePinho, R. A. (2000). A critical role for telomeres in suppressing and facilitating carcinogenesis. *Curr. Opin. Genet. Dev.*, *10(1)*, 39-46.
Harley, C. B., Futcher, A. B. & Greider, C. W. (1990). Telomeres shorten during ageing of human fibroblasts. *Nature*, *345*, 458-460.
Kim, N. W., Piatyszek, M. A., Prowse, K. R., Harley, C. B., West, M. D., Ho, P. L. C., Coviello, G. M., Wright, W. E., Weinrich, S. L. & Shay, J. W. (1994). Specific association of human telomerase activity with immortal cells and cancer. *Science*, *266(23)*, 2011-2015.
Counter, C. M., Avilion, A. A., LeFeuvre, C. E., Stewart, N. G., Greider, C. W., Harley, C. B. & Bacchetti, S. (1992). Telomere shortening associated with chromosome instability is arrested in immortal cells which express telomerase activity. *EMBO J.*, *11(5)*, 1921-1929.
Lingner, J., Hughes, T. R., Shevchenko, A., Mann, M., Lundblad, V. & Cech, T. R. (1997). Reverse transcriptase motifs in the catalytic subunit of telomerase. *Science*, *276(5312)*, 561-567.
Nakamura, T. M., Morin, G. B., Chapman, K. B., Weinrich, S. L., Andrews, W. H., Lingner, J., Harley, C. B. & Cech, T. R. (1997). Telomerase catalytic subunit homologs from fission yeast and human. *Science*, *277*, 955-959.
Bodnar, A. G., Ouellette, M., Frolkis, M., Holt, S. E., Chiu, C. P., Morin, G. B., Harley, C. B., Shay, J. W., Lichtsteiner, S. & Wright, W. E. (1998). Extension of life-span by introduction of telomerase into normal human cells [see comments]. *Science*, *279(5349)*, 349-352.
Maser, R. S. & DePinho, R. A. (2002). Connecting chromosomes, crisis, and cancer. *Science*, *297(5581)*, 565-569.
Hiyama, E. & Hiyama, K. (2002). Clinical utility of telomerase in cancer. *Oncogene*, *21(4)*, 643-649.
Dhaene, K., Van Marck, E. & Parwaresch, R. (2000). Telomeres, telomerase and cancer: An up-date. *Virchows Arch.*, *437(1)*, 1-16.
Kim, N. W. (1997). Clinical implications of telomerase in cancer. *Eur. J. Cancer*, *33(5)*, 781-786.

Shay, J. W. & Gazdar, A. F. (1997). Telomerase in the early detection of cancer. *J. Clin. Pathol.*, *50*, 106-109.

Hackett, J. A. & Greider, C. W. (2002). Balancing instability: Dual roles for telomerase and telomere dysfunction in tumorigenesis. *Oncogene*, *21(4)*, 619-626.

Blasco, M. A., Rizen, M., Greider, C. W. & Hanahan, D. (1996). Differential regulation of telomerase activity and telomerase rna during multi-stage tumorigenesis. *Nature Genet.*, *12*, 200-204.

Broccoli, D., Godley, L. A., Donehower, L. A., Varmus, H. E. & de Lange, T. (1996). Telomerase activation in mouse mammary tumors: Lack of detectable telomere shortening and evidence for regulation of telomerase rna with cell proliferation. *Mol. Cell Biol.*, *16(7)*, 3765-3772.

Hahn, W. C., Stewart, S. A., Brooks, M. W., York, S. G., Eaton, E., Kurachi, A., Beijersbergen, R. L., Knoll, J. H., Meyerson, M. & Weinberg, R. A. (1999). Inhibition of telomerase limits the growth of human cancer cells. *Nat. Med.*, *5(10)*, 1164-1170.

Harley, C. B., Vaziri, H., Counter, C. M. & Allsopp, R. C. (1992). The telomere hypothesis of cellular aging. *Exp. Gerontology*, *27*, 375-382.

Wright, W. E. & Shay, J. W. (1992). The two-stage mechanism controlling cellular senescence and immortalization. *Exp. Gerontology*, *27*, 383-389.

Vaziri, H. & Benchimol, S. (1998). Reconstitution of telomerase activity in normal human cells leads to elongation of telomeres and extended replicative life span. *Curr. Biol.*, *8(5)*, 279-282.

Yang, J., Chang, E., Cherry, A. M., Bangs, C. D., Oei, Y., Bodnar, A., Bronstein, A., Chiu, C. P. & Herron, G. S. (1999). Human endothelial cell life extension by telomerase expression. *J. Biol. Chem.*, *274(37)*, 26141-26148.

Hiyama, E. & Hiyama, K. (2003). Telomerase as tumor marker. *Cancer Lett.*, *194(2)*, 221-233.

Shay, J. W., Zou, Y., Hiyama, E. & Wright, W. E. (2001). Telomerase and cancer. *Hum. Mol. Genet.*, *10(7)*, 677-685.

Artandi, S. E., Chang, S., Lee, S. L., Alson, S., Gottlieb, G. J., Chin, L. & & DePinho, R. A. (2000). Telomere dysfunction promotes non-reciprocal translocations and epithelial cancers in mice. *Nature*, *406(6796)*, 641-645.

Rudolph, K. L., Millard, M., Bosenberg, M. W. & DePinho, R. A. (2001). Telomere dysfunction and evolution of intestinal carcinoma in mice and humans. *Nat. Genet.*, *28(2)*, 155-159.

Hashimoto, Y., Murakami, Y., Uemura, K., Hayashidani, Y., Sudo, T., Ohge, H., Fukuda, E., Shimamoto, F., Sueda, T. & Hiyama, E. (2008). Telomere shortening and telomerase expression during multistage carcinogenesis of intraductal papillary mucinous neoplasms of the pancreas. *J. Gastrointest. Surg.*, *12(1)*, 17-28; discussion 28-19.

Hashimoto, Y., Murakami, Y., Uemura, K., Hayashidani, Y., Sudo, T., Ohge, H., Sueda, T., Shimamoto, F. & Hiyama, E. (2008). Mixed ductal-endocrine carcinoma derived from intraductal papillary mucinous neoplasm (ipmn) of the pancreas identified by human telomerase reverse transcriptase (htert) expression. *J. Surg. Oncol.*, *97(5)*, 469-475.

Kleideiter, E., Friedrich, U., Mohring, A., Walker, S., Horing, E., Maier, K., Fritz, P., Thon, K. P. & Klotz, U. (2003). Telomerase activity in chronic inflammatory bowel disease. *Dig. Dis. Sci.*, *48(12)*, 2328-2332.

Shay, J. W. & Bacchetti, S. (1997). A survey of telomerase activity in human cancer. *Eur. J. Cancer*, *33*, 787-791.

Wright, W. E. & Shay, J. W. (1995). Time, telomeres and tumours: Is cellular senescence more than an anticancer mechanism? *Trends Cell Biol.*, *5(8)*, 293-297.

Hiyama, K., Hirai, Y., Kyoizumi, S., Akiyama, M., Hiyama, E., Piatyszek, M. A. & Shay, J. W. (1995). Activation of telomerase in human lymphocytes and hematopoietic progenitor cells. *J. Immunol.*, *155*, 3711-3715.

Hiyama, E., Hiyama, K., Tatsumoto, N., Kodama, T., Shay, J. W. & Yokoyama, T. (1996). Telomerase activity in human intestine. *Int. J. Oncol.*, *9*, 453-458.

Wright, W. E., Piatyszek, M. A., Rainey, W. E., Byrd, W. & Shay, J. W. (1996). Telomerase activity in human germline and embryonic tissues and cells. *Dev. Genet.*, *18*, 173-179.

Hiyama, E., Gollahan, L., Kataoka, T., Kuroi, K., Yokoyama, T., Gazdar, A. F., Hiyama, K., Piatyszek, M. A. & Shay, J. W. (1996). Telomerase activity in human breast tumors. *Journal of the National Cancer Institute*, *88*, 116-122.

Mathew, P., Valentine, M. B., Bowman, L. C., Rowe, S. T., Nash, M. B., Valentine, V. A., Cohn, S. L., Castleberry, R. P., Brodeur, G. M. & Look, A. T. (2001). Detection of mycn gene amplification in neuroblastoma by fluorescence in situ hybridization: A pediatric oncology group study. *Neoplasia*, *3(2)*, 105-109.

Iwao, T., Hiyama, E., Yokoyama, T., Tsuchida, A., Hiyama, K., Murakami, Y., Shimamoto, F., Shay, J. W. & Kajiyama, G. (1997). Telomerase activity for the preoperative diagnosis of pancreatic cancer. *Journal of the National Cancer Institute, 89(21)*, 1621-1623.

Feng, J., Funk, W. D., Wang, S. S., Weinrich, S. L., Avilion, A. A., Chiu, C. P., Adams, R. R., Chang, E., Allsopp, R. C., Yu, J., Le, S., West, M. D., Harley, C. B., Andrews, W. H., Greider, C. W. & Villeponteau, B. (1995). The rna component of human telomerase. *Science, 269*, 1236-1241.

Ohyashiki, K., Ohyashiki, J. H., Nishimaki, J., Toyama, K., Ebihara, Y., Kato, H., Wright, W. E. & Shay, J. W. (1997). Cytological detection of telomerase activity using an in situ telomeric repeat amplification protocol assay. *Cancer Res., 57(11)*, 2100-2103.

Koyanagi, K., Ozawa, S., Ando, N., Mukai, M., Kitagawa, Y., Ueda, M. & Kitajima, M. (2000). Telomerase activity as an indicator of malignant potential in iodine- nonreactive lesions of the esophagus. *Cancer, 88(7)*, 1524-1529.

Chou, S. J., Chen, C. M., Harn, H. J., Chen, C. J. & Liu, Y. C. (2001). In situ detection of htert mrna relates to ki-67 labeling index in papillary thyroid carcinoma. *J. Surg. Res., 99(1)*, 75-83.

Kumaki, F., Kawai, T., Hiroi, S., Shinomiya, N., Ozeki, Y., Ferrans, V. J. & Torikata, C. (2001). Telomerase activity and expression of human telomerase rna component and human telomerase reverse transcriptase in lung carcinomas. *Hum. Pathol., 32(2)*, 188-195.

Hiyama, E., Hiyama, K., Shay, J. W. & Yokoyama, T. (2001). Immunhistochemical detection of telomerase (htert) protein in human cancer tissues and a subset of cells in normal tissues. *Neoplasia, 3(1)*, 17-26.

Kumaki, F., Takeda, K., Yu, Z. X., Moss, J. & Ferrans, V. J. (2002). Expression of human telomerase reverse transcriptase in lymphangioleiomyomatosis. *Am. J. Respir. Crit. Care Med., 166(2)*, 187-191.

Frost, M., Bobak, J. B., Gianani, R., Kim, N., Weinrich, S., Spalding, D. C., Cass, L. G., Thompson, L. C., Enomoto, T., Uribe-Lopez, D. & Shroyer, K. R. (2000). Localization of telomerase htert protein and htr in benign mucosa, dysplasia, and squamous cell carcinoma of the cervix. *Am. J. Clin. Pathol., 114(5)*, 726-734.

Miyazu, Y. M., Miyazawa, T., Hiyama, K., Kurimoto, N., Iwamoto, Y., Matsuura, H., Kanoh, K., Kohno, N., Nishiyama, M. & Hiyama, E. (2005). Telomerase expression in noncancerous bronchial epithelia is a possible marker of early development of lung cancer. *Cancer Res.*, *65(21)*, 9623-9627.

Slaughter, D. P., Southwick, H. W. & Smejkal, W. (1953). Field cancerization in oral stratified squamous epithelium; clinical implications of multicentric origin. *Cancer*, *6(5)*, 963-968.

Going, J. J., Fletcher-Monaghan, A. J., Neilson, L., Wisman, B. A., van der Zee, A., Stuart, R. C. & Keith, W. N. (2004). Zoning of mucosal phenotype, dysplasia, and telomerase activity measured by telomerase repeat assay protocol in barrett's esophagus. *Neoplasia*, *6(1)*, 85-92.

Luzar, B., Poljak, M., Marin, I. J., Eberlinc, A., Klopcic, U. & Gale, N. (2004). Human telomerase catalytic subunit gene re-expression is an early event in oral carcinogenesis. *Histopathology*, *45(1)*, 13-19.

Oikonomou, P., Mademtzis, I., Messinis, I. & Tsezou, A. (2006). Quantitative determination of human telomerase reverse transcriptase messenger rna expression in premalignant cervical lesions and correlation with human papillomavirus load. *Hum. Pathol.*, *37(2)*, 135-142.

Sen, S., Reddy, V. G., Guleria, R., Jain, S. K., Kapila, K. & Singh, N. (2002). Telomerase--a potential molecular marker of lung and cervical cancer. *Clin. Chem. Lab. Med.*, *40(10)*, 994-1001.

Sakr, W. A. & Partin, A. W. (2001). Histological markers of risk and the role of high-grade prostatic intraepithelial neoplasia. *Urology*, *57(4 Suppl 1)*, 115-120.

Chadeneau, C., Hay, K., Hirte, H. W., Gallinger, S. & Bacchetti, S. (1995). Telomerase activity associated with acquisition of malignancy in human colorectal cancer. *Cancer Research*, *55*, 2533-2536.

Hiyama, E., Yokoyama, T., Tatsumoto, N., Hiyama, K., Imamura, Y., Murakami, Y., Kodama, T., Piatyszek, M. A., Shay, J. W. & Matsuura, Y. (1995). Telomerase activity in gastric cancer. *Cancer Research*, *55(8)*, 3258-3262.

Kido, A., Schneider-Stock, R., Hauptmann, K. & Roessner, A. (2003). Telomerase activity in juxtacortical and conventional high-grade osteosarcomas: Correlation with grade, proliferative activity and clinical response to chemotherapy. *Cancer Lett.*, *196(1)*, 109-115.

Marchetti, A., Bertacca, G., Buttitta, F., Chella, A., Quattrocolo, G., Angeletti, C. A. & Bevilacqua, G. (1999). Telomerase activity as a prognostic indicator in stage i non-small cell lung cancer. *Clin. Cancer Res.*, *5(8)*, 2077-2081.

Tatsumoto, N., Hiyama, E., Murakami, Y., Imamura, Y., Shay, J. W., Matsuura, Y. & Yokoyama, T. (2000). High telomerase activity is an independent prognostic indicator of poor outcome in colorectal cancer. *Clinical Cancer Research*, 6, 2696-2701.

Huang, F., Kanno, H., Yamamoto, I., Lin, Y. & Kubota, Y. (1999). Correlation of clinical features and telomerase activity in human gliomas. *J. Neurooncol.*, *43(2)*, 137-142.

Nakatani, K., Yoshimi, N., Mori, H., Yoshimura, S., Sakai, H., Shinoda, J. & Sakai, N. (1997). The significant role of telomerase activity in human brain tumors. *Cancer*, *80(3)*, 471-476.

Maes, L., Kalala, J. P., Cornelissen, M. & de Ridder, L. (2007). Progression of astrocytomas and meningiomas: An evaluation in vitro. *Cell Prolif*, *40(1)*, 14-23.

Tabori, U., Ma, J., Carter, M., Zielenska, M., Rutka, J., Bouffet, E., Bartels, U., Malkin, D. & Hawkins, C. (2006). Human telomere reverse transcriptase expression predicts progression and survival in pediatric intracranial ependymoma. *J. Clin. Oncol.*, *24(10)*, 1522-1528.

Falchetti, M. L., Pierconti, F., Casalbore, P., Maggiano, N., Levi, A., Larocca, L. M. & Pallini, R. (2003). Glioblastoma induces vascular endothelial cells to express telomerase in vitro. *Cancer Res.*, *63(13)*, 3750-3754.

Pallini, R., Pierconti, F., Falchetti, M. L., D'Arcangelo, D., Fernandez, E., Maira, G., D'Ambrosio, E. & Larocca, L. M. (2001). Evidence for telomerase involvement in the angiogenesis of astrocytic tumors: Expression of human telomerase reverse transcriptase messenger rna by vascular endothelial cells. *J. Neurosurg.*, *94(6)*, 961-971.

Yoshino, A., Katayama, Y., Fukushima, T., Watanabe, T., Komine, C., Yokoyama, T., Kusama, K. & Moro, I. (2003). Telomerase activity in pituitary adenomas: Significance of telomerase expression in predicting pituitary adenoma recurrence. *J. Neurooncol.*, *63(2)*, 155-162.

Hiyama, E., Hiyama, K., Ohtsu, K., Yamaoka, H., Ichikawa, T., Shay, J. W. & Yokoyama, T. (1997). Telomerase activity in neuroblastoma: Is it a prognostic indicator of clinical behavior? *Eurepean Journal of Cancer*, *33(12)*, 1932-1936.

Hiyama, E., Hiyama, K., Yokoyama, T., Matsuura, Y., Piatyszek, M. A. & Shay, J. W. (1995). Correlating telomerase activity levels with human neuroblastoma outcomes. *Nature Medicine*, *1(3)*, 249-255.

Krams, M., Hero, B., Berthold, F., Parwaresch, R., Harms, D. & Rudolph, P. (2003). Full-length telomerase reverse transcriptase messenger rna is an independent prognostic factor in neuroblastoma. *Am. J. Pathol.*, *162(3)*, 1019-1026.

Poremba, C., Willenbring, H., Hero, B., Christiansen, H., Schafer, K. L., Brinkschmidt, C., Jurgens, H., Bocker, W. & Dockhorn-Dworniczak, B. (1999). Telomerase activity distinguishes between neuroblastomas with good and poor prognosis. *Ann. Oncol., 10(6)*, 715-721.

Streutker, C. J., Thorner, P., Fabricius, N., Weitzman, S. & Zielenska, M. (2001). Telomerase activity as a prognostic factor in neuroblastomas. *Pediatr. Dev. Pathol., 4(1)*, 62-67.

Nozaki, C., Horibe, K., Iwata, H., Ishiguro, Y., Hamaguchi, M. & Takahashi, M. (2000). Prognostic impact of telomerase activity in patients with neuroblastoma. *Int. J. Oncol., 17(2)*, 341-345.

Hiyama, E., Yamaoka, H., Matsunaga, T., Hayashi, Y., Ando, H., Suita, S., Horie, H., Kaneko, M., Sasaki, F., Hashizume, K., Nakagawara, A., Ohnuma, N. & Yokoyama, T. (2004). High expression of telomerase is an independent prognostic indicator of poor outcome in hepatoblastoma. *Br. J. Cancer, 91(5)*, 972-979.

Kleideiter, E., Schwab, M., Friedrich, U., Koscielniak, E., Schafer, B. W. & Klotz, U. (2003). Telomerase activity in cell lines of pediatric soft tissue sarcomas. *Pediatr. Res., 54(5)*, 718-723.

Smith, D. L., Soria, J. C., Morat, L., Yang, Q., Sabatier, L., Liu, D. D., Nemr, R. A., Rashid, A. & Vauthey, J. N. (2004). Human telomerase reverse transcriptase (htert) and ki-67 are better predictors of survival than established clinical indicators in patients undergoing curative hepatic resection for colorectal metastases. *Ann. Surg. Oncol., 11(1)*, 45-51.

Clark, G. M., Osborne, C. K., Levitt, D., Wu, F. & Kim, N. W. (1997). Telomerase activity and survival of patients with node-positive breast cancer. *J. Natl. Cancer Inst., 89(24)*, 1874-1881.

Elkak, A., Mokbel, R., Wilson, C., Jiang, W. G., Newbold, R. F. & Mokbel, K. (2006). Htert mrna expression is associated with a poor clinical outcome in human breast cancer. *Anticancer Res., 26(6C)*, 4901-4904.

Kammori, M., Izumiyama, N., Hashimoto, M., Nakamura, K., Okano, T., Kurabayashi, R., Naoki, H., Honma, N., Ogawa, T., Kaminishi, M. & Takubo, K. (2005). Expression of human telomerase reverse transcriptase gene and protein, and of estrogen and progesterone receptors, in breast tumors: Preliminary data from neo-adjuvant chemotherapy. *Int. J. Oncol., 27(5)*, 1257-1263.

Haugen, B. R., Nawaz, S., Markham, N., Hashizumi, T., Shroyer, A. L., Werness, B. & Shroyer, K. R. (1997). Telomerase activity in benign and malignant thyroid tumors. *Thyroid, 7(3)*, 337-342.

Saji, M., Westra, W. H., Chen, H., Umbricht, C. B., Tuttle, R. M., Box, M. F., Udelsman, R., Sukumar, S. & Zeiger, M. A. (1997). Telomerase activity in the differential diagnosis of papillary carcinoma of the thyroid. *Surgery, 122(6)*, 1137-1140.

Tomoda, R., Seto, M., Tsumuki, H., Iida, K., Yamazaki, T., Sonoda, J., Matsumine, A. & Uchida, A. (2002). Telomerase activity and human telomerase reverse transcriptase mrna expression are correlated with clinical aggressiveness in soft tissue tumors. *Cancer, 95(5)*, 1127-1133.

Sumida, T., Sogawa, K., Hamakawa, H., Sugita, A., Tanioka, H. & Ueda, N. (1998). Detection of telomerase activity in oral lesions. *J. Oral pathol. Med., 27(3)*, 111-115.

Mao, L., El-Naggar, A. K., Fan, Y. H., Lee, J. S., Lippman, S. M., Kayser, S., Lotan, R. & Hong, W. K. (1996). Telomerase activity in head and neck squamous cell carcinoma and adjacent tissues. *Cancer Res., 56(24)*, 5600-5604.

Sumida, T., Hamakawa, H., Sogawa, K., Sugita, A., Tanioka, H. & Ueda, N. (1999). Telomerase components as a diagnostic tool in human oral lesions. *Int. J. Cancer, 80(1)*, 1-4.

Califano, J., Ahrendt, S. A., Meininger, G., Westra, W. H., Koch, W. H. & Sidransky, g. (1998). Detection of telomerase activity in oral renses from head and neck squamous cell carcinoma. *Cancer Res., 56(24)*, 5720-5722.

Capkova, L., Kalinova, M., Krskova, L., Kodetova, D., Petrik, F., Trefny, M., Musil, J. & Kodet, R. (2007). Loss of heterozygosity and human telomerase reverse transcriptase (htert) expression in bronchial mucosa of heavy smokers. *Cancer, 109(11)*, 2299-2307.

Sen, S., Reddy, V. G., Khanna, N., Guleria, R., Kapila, K. & Singh, N. (2001). A comparative study of telomerase activity in sputum, bronchial washing and biopsy specimens of lung cancer. *Lung Cancer, 33(1)*, 41-49.

Haruta, Y., Hiyama, K., Ishioka, S., Hozawa, S., Hiroaki, M. & Yamakido, M. (1999). Activation of telomerase is induced by a natural antigen in allergen-specific memory t lymphocytes in broncheal asthma. *Biochem. Biophys. Res. Commun., 259(3)*, 617-623.

Hiyama, K., Ishioka, S., Shay, J. W., Taooka, Y., Maeda, A., Isobe, T., Hiyama, E., Maeda, H. & Yamakido, M. (1998). Telomerase activity as a novel marker of lung cancer and immune-associated lung diseases. *International Journal of Molecular Medicine, 1*, 545-549.

Yoshida, K., Sugino, T., Goodison, S., Warren, B. F., Nolan, D., Wadsworth, S., Mortensen, N. J., Toge, T., Tahara, E. & Tarin, D. (1997). Detection of telomerase activity in exfoliated cancer cells in colonic luminal washings and its related clinical implications. *Br. J. Cancer, 75(4)*, 548-553.

Hiyama, E., Kodama, T., Sinbara, K., Iwao, T., Itoh, M., Hiyama, K., Shay, J. W., Matsuura, Y. & Yokoyama, T. (1997). Telomerase activity is detected in pancreatic cancer but not in benign tumors. *Cancer Research, 57*, 326-331.

Morales, C. P., Burdick, J. S., Saboorian, M. H., Wright, W. E. & Shay, J. W. (1998). In situ hybridization for telomerase rna in routine cytologic brushings for the diagnosis of pancreaticobiliary malignancies. *Gastrointest Endosc, 48(4)*, 402-405.

Suehara, N., Mizumoto, K., Tanaka, M., Niiyama, H., Yokohata, K., Yominaga, Y., Shimura, H., Muta, T. & Hamasaki, N. (1997). Telomerase activity in pancreatic juice differentiates ductal carcinoma from adenoma and pancreatitis. *Clin. Cancer Res., 3*, 2479-2483.

Inoue, H., Tsuchida, A., Kawasaki, Y., Fujimoto, Y., Yamasaki, S. & Kajiyama, G. (2001). Preoperative diagnosis of intraductal papillary-mucinous tumors of the pancreas with attention to telomerase activity. *Cancer, 91(1)*, 35-41.

Uemura, K., Hiyama, E., Murakami, Y., Kanehiro, T., Ohge, H., Sueda, T. & Yokoyama, T. (2003). Comparative analysis of k-ras point mutation, telomerase activity, and p53 overexpression in pancreatic tumours. *Oncol. Rep., 10(2)*, 277-283.

Hashimoto, Y., Murakami, Y., Uemura, K., Hayashidani, Y., Sudo, T., Ohge, H., Fukuda, E., Sueda, T. & Hiyama, E. (2008). Detection of human telomerase reverse transcriptase (htert) expression in tissue and pancreatic juice from pancreatic cancer. *Surgery, 143(1)*, 113-125.

Itoi, T., Ohyashiki, K., Yahata, N., Shinohara, Y., Takei, K., Takeda, K., Nagao, K., Hisatomi, H., Ebihara, Y., Shay, J. W. & Saito, T. (1999). Detection of telomerase activity in exfoliated cancer cells obtained from bile. *Int. J. Oncol., 15(6)*, 1061-1067.

Itoi, T., Shinohara, Y., Takeda, K., Nakamura, K., Shimizu, M., Ohyashiki, K., Hisatomi, H., Nakano, H. & Moriyasu, F. (2001). Detection of telomerase reverse transcriptase mrna in biopsy specimens and bile for diagnosis of biliary tract cancers. *Int. J. Mol. Med., 7(3)*, 281-287.

Shukla, V. K., Chauhan, V. S. & Kumar, M. (2006). Telomerase activation--one step on the road to carcinoma of the gall bladder. *Anticancer Res.*, *26(6C)*, 4761-4766.

Ramakumar, S., Bhuiyan, J., Besse, J. A., Roberts, S. G., Wollan, P. C., Blute, M. L. & O'Kane, D. J. (1999). Comparison of screening methods in the detection of bladder cancer. *J. Urol.*, *161(2)*, 388-394.

Fukui, T., Nonomura, N., Tokizane, T., Sato, E., Ono, Y., Harada, Y., Nishimura, K., Takahara, S. & Okuyama, A. (2001). Clinical evaluation of human telomerase catalytic subunit in bladder washings from patients with bladder cancer. *Mol. Urol.*, *5(1)*, 19-23.

Ito, H., Kyo, S., Kanaya, T., Takakura, M., Koshida, K., Namiki, M. & Inoue, M. (1998). Detection of human telomerase reverse transcriptase messenger rna in voided urine samples as a useful diagnostic tool for bladder cancer. *Clin. Cancer Res.*, *4(11)*, 2807-2810.

Eissa, S., Swellam, M., Ali-Labib, R., Mansour, A., El-Malt, O. & Tash, F. M. (2007). Detection of telomerase in urine by 3 methods: Evaluation of diagnostic accuracy for bladder cancer. *J. Urol.*, *178(3 Pt 1)*, 1068-1072.

Lee, D. H., Yang, S. C., Hong, S. J., Chung, B. H. & Kim, I. Y. (1998). Telomerase: A potential marker of bladder transitional cell carcinoma in bladder washes. *Clin. Cancer Res.*, *4(3)*, 535-538.

Longchampt, E., Lebret, T., Molinie, V., Bieche, I., Botto, H. & Lidereau, R. (2003). Detection of telomerase status by semiquantitative and in situ assays, and by real-time reverse transcription-polymerase chain reaction (telomerase reverse transcriptase) assay in bladder carcinomas. *BJU Int.*, *91(6)*, 567-572.

Meid, F. H., Gygi, C. M., Leisinger, H. J., Bosman, F. T. & Benhattar, J. (2001). The use of telomerase activity for the detection of prostatic cancer cells after prostatic massage. *J. Urol.*, *165(5)*, 1802-1805.

Vicentini, C., Gravina, G. L., Angelucci, A., Pascale, E., D'Ambrosio, E., Muzi, P., Di Leonardo, G., Fileni, A., Tubaro, A., Festuccia, C. & Bologna, M. (2004). Detection of telomerase activity in prostate massage samples improves differentiating prostate cancer from benign prostatic hyperplasia. *J. Cancer Res. Clin. Oncol.*, *130(4)*, 217-221.

Wang, Z., Ramin, S. A., Tsai, C., Lui, P., Ruckle, H. C., Beltz, R. E. & Sands, J. F. (2001). Telomerase activity in prostate sextant needle cores from radical prostatectomy specimens. *Urol. Oncol.*, *6(2)*, 57-62.

Wang, Z., Ramin, S. A., Tsai, C., Lui, P., Ruckle, H. C., Beltz, R. E., Sands, J. F. & Slattery, C. W. (2002). Evaluation of pcr-elisa for determination of telomerase activity in prostate needle biopsy and prostatic fluid specimens. *Urol. Oncol., 7(5)*, 199-205.

Umbricht, C. B., Saji, M., Westra, W. H., Udelsman, R., Zeiger, M. A. & Sukumar, S. (1997). Telomerase activity: A marker to distinguish follicular thyroid adenoma from carcinoma. *Cancer Res., 57(11)*, 2144-2147.

Zeiger, M. A., Smallridge, R. C., Clark, D. P., Liang, C. K., Carty, S. E., Watson, C. G., Udelsman, R. & Saji, M. (1999). Human telomerase reverse transcriptase (htert) gene expression in fna samples from thyroid neoplasms. *Surgery, 126(6)*, 1195-1198; discussion 1198-1199.

Mora, J. & Lerma, E. (2004). Telomerase activity in thyroid fine needle aspirates. *Acta. Cytol., 48(6)*, 818-824.

Trulsson, L. M., Velin, A. K., Herder, A., Soderkvist, P., Ruter, A. & Smeds, S. (2003). Telomerase activity in surgical specimens and fine-needle aspiration biopsies from hyperplastic and neoplastic human thyroid tissues. *Am. J. Surg., 186(1)*, 83-88.

Kammori, M., Nakamura, K., Hashimoto, M., Ogawa, T., Kaminishi, M. & Takubo, K. (2003). Clinical application of human telomerase reverse transcriptase gene expression in thyroid follicular tumors by fine-needle aspirations using in situ hybridization. *Int. J. Oncol., 22(5)*, 985-991.

Umbricht, C. B., Sherman, M. E., Dome, J., Carey, L. A., Marks, J., Kim, N. & Sukumar, S. (1999). Telomerase activity in ductal carcinoma in situ and invasive breast cancer. *Oncogene, 18(22)*, 3407-3414.

Bieche, I., Nogues, C., Paradis, V., Olivi, M., Bedossa, P., Lidereau, R. & Vidaud, M. (2000). Quantitation of htert gene expression in sporadic breast tumors with a real-time reverse transcription-polymerase chain reaction assay. *Clin. Cancer Res., 6(2)*, 452-459.

Pearson, A. S., Gollahon, L. S., O'Neal, N. C., Saboorian, H., Shay, J. W. & Fahey, T. r. (1998). Detection of telomerase activity in breast masses by fine-needle aspiration. *Ann. Surg. Oncol., 5(2)*, 186-193.

Kirkpatrick, K. L., Ogunkolade, W., Elkak, A. E., Bustin, S., Jenkins, P., Ghilchick, M., Newbold, R. F. & Mokbel, K. (2003). Htert expression in human breast cancer and non-cancerous breast tissue: Correlation with tumour stage and c-myc expression. *Breast Cancer Res. Treat, 77(3)*, 277-284.

Hines, W. C., Fajardo, A. M., Joste, N. E., Bisoffi, M. & Griffith, J. K. (2005). Quantitative and spatial measurements of telomerase reverse transcriptase expression within normal and malignant human breast tissues. *Mol. Cancer Res., 3(9)*, 503-509.

Hiyama, E., Saeki, T., Hiyama, K., Takashima, S., Shay, J. W., Matsuura, Y. & Yokoyama, T. (2000). Telomerase activity as a marker of breast cancer in fine needle aspirated samples. *Cancer Cytopathol.*, *90(4)*, 235-238.

Mokbel, K., Williams, N. J., Leris, A. C. & Kouriefs, C. (1999). Telomerase activity in fine-needle aspirates of breast lesions. *J. Clin Oncol.*, *17(12)*, 3856-3860.

Poremba, C., Shroyer, K. R., Frost, M., Diallo, R., Fogt, F., Schafer, K. L., Burger, H., Shroyer, A. L., Dockhom-Dwomiczak, B. & Bocker, W. (1999). Telomerase is a highly sensitive and specific molecular marker in fine-needle aspirates in breast lesions. *J. Clin Oncol.*, *17(7)*, 2020-2026.

Zhang, S., Dong, M., Teng, X. & Chen, T. (2001). Quantitative assay of telomerase activity in head and neck squamous cell carcinoma and other tissues. *Arch. Otolaryngol. Head Neck Surg.*, *127(5)*, 581-585.

Barclay, J. Y., Morris, A. & Nwokolo, C. U. (2005). Telomerase, htert and splice variants in barrett's oesophagus and oesophageal adenocarcinoma. *Eur. J. Gastroenterol. Hepatol.*, *17(2)*, 221-227.

Yang, S. M., Fang, D. C., Luo, Y. H., Lu, R., Battle, P. D. & Liu, W. W. (2001). Alterations of telomerase activity and terminal restriction fragment in gastric cancer and its premalignant lesions. *J. Gastroenterol. Hepatol.*, *16(8)*, 876-882.

Nagao, K., Tomimatsu, M., Endo, H., Hisatomi, H. & Hikiji, K. (1999). Telomerase reverse transcriptase mrna expression and telomerase activity in hepatocellular carcinoma. *J. Gastroenterol.*, *34(1)*, 83-87.

Nakashio, R., Kitamoto, M., Tahara, H., Nakanishi, T., Ide, T. & Kajiyama, G. (1997). Significance of telomerase activity in the diagnosis of small differentiated hepatocellular carcinoma. *Int. J. Cancer*, *74(2)*, 141-147.

Shimada, M., Hasegawa, H., Gion, T., Utsunomiya, T., Shirabe, K., Takenaka, K., Otsuka, T., Maehara, Y. & Sugimachi, K. (2000). The role of telomerase activity in hepatocellular carcinoma. *Am. J. Gastroenterol.*, *95(3)*, 748-752.

Jarboe, E. A., Liaw, K. L., Thompson, L. C., Heinz, D. E., Baker, P. L., McGregor, J. A., Dunn, T., Woods, J. E. & Shroyer, K. R. (2002). Analysis of telomerase as a diagnostic biomarker of cervical dysplasia and carcinoma. *Oncogene*, *21(4)*, 664-673.

Wisman, G. B., Hollema, H., de jong, S., ter Schegget, J., Tjong-A-Hung, S. P., Ruiters, M. H., Krans, M., de Vries, E. G. & van der Zee, A. G. (1998). Telomerase activity as a biomarker for (pre)neoplastic cervical disease in scrapings and frozen sections from patients with abnormal cervical smear. *J. Clin. Oncol.*, *16(6)*, 2238-2245.

Zheng, P. S., Iwasaka, T., Zhang, Z. M., Pater, A. & Sugimori, H. (2000). Telomerase activity in papanicolaou smear-negative exfoliated cervical cells and its association with lesions and oncogenic human papillomaviruses. *Gynecol. Oncol.*, *77(3)*, 394-398.

Triginelli, S. A., Silva-Filho, A. L., Traiman, P., Silva, F. M., Chaves-Dias, M. C., Oliveira, G. C. & Cunha-Melo, J. R. (2006). Telomerase activity in the vaginal margins of radical hysterectomy in patients with carcinoma of the cervix: Correlation with histology and human papillomavirus. *Int. J. Gynecol. Cancer*, *16(3)*, 1283-1288.

Reesink-Peters, N., Helder, M. N., Wisman, G. B., Knol, A. J., Koopmans, S., Boezen, H. M., Schuuring, E., Hollema, H., de Vries, E. G., de Jong, S. & van der Zee, A. G. (2003). Detection of telomerase, its components, and human papillomavirus in cervical scrapings as a tool for triage in women with cervical dysplasia. *J. Clin. Pathol.*, *56(1)*, 31-35.

Parris, C. N., Jezzard, S., Silver, A., MacKie, R., McGregor, J. M. & Newbold, R. F. (1999). Telomerase activity in melanoma and non-melanoma skin cancer. *Br. J. Cancer*, *79(1)*, 47-53.

Zygouris, P., Tsiambas, E., Tiniakos, D., Karameris, A., Athanassiou, A. E., Kittas, C. & Kyroudi, A. (2007). Evaluation of combined h-tert, bcl-2, and caspases 3 and 8 expression in cutaneous malignant melanoma based on tissue microarrays and computerized image analysis. *J. Buon.*, *12(4)*, 513-519.

Fullen, D. R., Zhu, W., Thomas, D. & Su, L. D. (2005). Htert expression in melanocytic lesions: An immunohistochemical study on paraffin-embedded tissue. *J. Cutan. Pathol.*, *32(10)*, 680-684.

Schneider-Stock, R., Rys, J., Jaeger, V., Niezabitowski, A., Kruczak, A., Sokolowski, A. & Roessner, A. (1999). Prognostic significance of telomerase activity in soft tissue sarcomas. *Int. J. Oncol.*, *15(4)*, 775-780.

Sabah, M., Cummins, R., Leader, M. & Kay, E. (2006). Immunohistochemical detection of htert protein in soft tissue sarcomas: Correlation with tumor grade. *Appl. Immunohistochem. Mol. Morphol.*, *14(2)*, 198-202.

Terasaki, T., Kyo, S., Takakura, M., Maida, Y., Tsuchiya, H., Tomita, K. & Inoue, M. (2004). Analysis of telomerase activity and telomere length in bone and soft tissue tumors. *Oncol. Rep.*, *11(6)*, 1307-1311.

Yan, P., Benhattar, J., Coindre, J. M. & Guillou, L. (2002). Telomerase activity and htert mrna expression can be heterogeneous and does not correlate with telomere length in soft tissue sarcomas. *Int. J. Cancer*, *98(6)*, 851-856.

Guilleret, I., Yan, P., Guillou, L., Braunschweig, R., Coindre, J. M. & Benhattar, J. (2002). The human telomerase rna gene (hterc) is regulated during carcinogenesis but is not dependent on DNA methylation. *Carcinogenesis*, *23(12)*, 2025-2030.

Umehara, N., Ozaki, T., Sugihara, S., Kunisada, T., Morimoto, Y., Kawai, A., Nishida, K., Yoshida, A., Murakami, T. & Inoue, H. (2004). Influence of telomerase activity on bone and soft tissue tumors. *J. Cancer Res. Clin. Oncol.*, *130(7)*, 411-416.

Cairney, C. J., Hoare, S. F., Daidone, M. G., Zaffaroni, N. & Keith, W. N. (2008). High level of telomerase rna gene expression is associated with chromatin modification, the alt phenotype and poor prognosis in liposarcoma. *Br. J. Cancer*, *98(8)*, 1467-1474.

Braunschweig, R., Yan, P., Guilleret, I., Delacretaz, F., Bosman, F. T., Mihaescu, A. & Benhattar, J. (2001). Detection of malignant effusions: Comparison of a telomerase assay and cytologic examination. *Diagn. Cytopathol.*, *24(3)*, 174-180.

Dejmek, A., Yahata, N., Ohyashiki, K., Ebihara, Y., Kakihana, M., Hirano, T., Kawate, N. & Kato, H. (2001). In situ telomerase activity in pleural effusions: A promising marker for malignancy. *Diagn. Cytopathol.*, *24(1)*, 11-15.

Wallace, M. B., Block, M., Hoffman, B. J., Hawes, R. H., Silvestri, G., Reed, C. E., Mitas, M., Ravenel, J., Fraig, M., Miller, S., Jones, E. T. & Boylan, A. (2003). Detection of telomerase expression in mediastinal lymph nodes of patients with lung cancer. *Am. J. Respir. Crit. Care Med.*, *167(12)*, 1670-1675.

Yang, C. T., Lee, M. H., Lan, R. S. & Chen, J. K. (1998). Telomerase activity in pleural effusions: Diagnostic significance. *J. Clin. Oncol.*, *16(2)*, 567-573.

Tangkijvanich, P., Tresukosol, D., Sampatanukul, P., Sakdikul, S., Voravud, N., Mahachai, V. & Mutirangura, A. (1999). Telomerase assay for differentiating between malignancy-related and nonmalignant ascites. *Clin. Cancer Res.*, *5(9)*, 2470-2475.

Duggan, B. D., Wan, M., Yu, M. C., Roman, L. D., Muderspach, L. I., Delgadillo, E., Li, W. Z., Martin, S. E. & Dubeau, L. (1998). Detection of ovarian cancer cells: Comparison of a telomerase assay and cytologic examination. *J. Natl. Cancer Inst.*, *90(3)*, 238-242.

Gauthier, L. R., Granotier, C., Soria, J. C., Faivre, S., Boige, V., Raymond, E. & Boussin, F. D. (2001). Detection of circulating carcinoma cells by telomerase activity. *Br. J. Cancer*, *84(5)*, 631-635.

Chen, X. Q., Bonnefoi, H., Pelte, M. F., Lyautey, J., Lederrey, C., Movarekhi, S., Schaeffer, P., Mulcahy, H. E., Meyer, P., Stroun, M. & Anker, P. (2000). Telomerase rna as a detection marker in the serum of breast cancer patients. *Clin. Cancer Res.*, *6(10)*, 3823-3826.

Miura, N., Maeda, Y., Kanbe, T., Yazama, H., Takeda, Y., Sato, R., Tsukamoto, T., Sato, E., Marumoto, A., Harada, T., Sano, A., Kishimoto, Y., Hirooka, Y., Murawaki, Y., Hasegawa, J. & Shiota, G. (2005). Serum human telomerase reverse transcriptase messenger rna as a novel tumor marker for hepatocellular carcinoma. *Clin. Cancer Res.*, *11(9)*, 3205-3209.

Miura, N., Maruyama, S., Oyama, K., Horie, Y., Kohno, M., Noma, E., Sakaguchi, S., Nagashima, M., Kudo, M., Kishimoto, Y., Kawasaki, H., Hasegawa, J. & Shiota, G. (2007). Development of a novel assay to quantify serum human telomerase reverse transcriptase messenger rna and its significance as a tumor marker for hepatocellular carcinoma. *Oncology*, *72 Suppl 1*, 45-51.

Orlando, C., Gelmini, S., Selli, C. & Pazzagli, M. (2001). Telomerase in urological malignancy. *J. Urol.*, *166(2)*, 666-673.

Harada, K., Kurisu, K., Arita, K., Sadatomo, T., Tahara, H., Tahara, E., Ide, T. & Uozumi, T. (1999). Telomerase activity in central nervous system malignant lymphoma. *Cancer*, *86(6)*, 1050-1055.

Ogawa, Y., Nishioka, A., Hamada, N., Terashima, M., Inomata, T., Yoshida, S., Seguchi, H. & Kishimoto, S. (1998). Changes in telomerase activity of advanced cancers of oral cavity and oropharynx during radiation therapy: Correlation with clinical outcome. *Int. J. Mol. Med.*, *2(3)*, 301-307.

Patel, M. M., Parekh, L. J., Jha, F. P., Sainger, R. N., Patel, J. B., Patel, D. D., Shah, P. M. & Patel, P. S. (2002). Clinical usefulness of telomerase activation and telomere length in head and neck cancer. *Head Neck*, *24(12)*, 1060-1067.

Gonzalez-Quevedo, R., Iniesta, P., Moran, A., de Juan, C., Sanchez-Pernaute, A., Fernandez, C., Torres, A., Diaz-Rubio, E., Balibrea, J. L. & Benito, M. (2002). Cooperative role of telomerase activity and p16 expression in the prognosis of non-small-cell lung cancer. *J. Clin. Oncol.*, *20(1)*, 254-262.

Hirashima, T., Komiya, T., Nitta, T., Takada, Y., Kobayashi, M., Masuda, N., Matui, K., Takada, M., Kikui, M., Yasumitu, T., Ohno, A., Nakagawa, K., Fukuoka, M. & Kawase, I. (2000). Prognostic significance of telomeric repeat length alterations in pathological stage i-iiia non-small cell lung cancer. *Anticancer Res., 20(3B)*, 2181-2187.

Taga, S., Osaki, T., Ohgami, A., Imoto, H. & Yasumoto, K. (1999). Prognostic impact of telomerase activity in non-small cell lung cancers. *Ann. Surg., 230(5)*, 715-720.

Hara, H., Yamashita, K., Shinada, J., Yoshimura, H. & Kameya, T. (2001). Clinicopathologic significance of telomerase activity and htert mrna expression in non-small cell lung cancer. *Lung Cancer, 34(2)*, 219-226.

Soria, J. C., Moon, C., Wang, L., Hittelman, W. N., Jang, S. J., Sun, S. Y., Lee, J. J., Liu, D., Kurie, J. M., Morice, R. C., Lee, J. S., Hong, W. K. & Mao, L. (2001). Effects of n-(4-hydroxyphenyl)retinamide on htert expression in the bronchial epithelium of cigarette smokers. *J. Natl. Cancer Inst., 93(16)*, 1257-1263.

Carey, L. A., Kim, N. W., Goodman, S., Marks, J., Henderson, G., Umbricht, C. B., Dome, J. S., Dooley, W., Amshey, S. R. & Sukumar, S. (1999). Telomerase activity and prognosis in primary breast cancers. *J. Clin. Oncol., 17(10)*, 3075-3081.

Okayasu, I., Osakabe, T., Fujiwara, M., Fukuda, H., Kato, M. & Oshimura, M. (1997). Significant correlation of telomerase activity in thyroid papillary carcinomas with cell differentiation, proliferation and extrathyroidal extension. *Jpn. J. Cancer Res., 88(10)*, 965-970.

Ito, M., Liu, Y., Yang, Z., Nguyen, J., Liang, F., Morris, R. J. & Cotsarelis, G. (2005). Stem cells in the hair follicle bulge contribute to wound repair but not to homeostasis of the epidermis. *Nat. Med., 11(12)*, 1351-1354.

Kakeji, Y., Maehara, Y., Koga, T., Shibahara, K., Kabashima, A., Tokunaga, E. & Sugimachi, K. (2001). Gastric cancer with high telomerase activity shows rapid development and invasiveness. *Oncol. Rep., 8(1)*, 107-110.

Usselmann, B., Newbold, M., Morris, A. G. & Nwokolo, C. U. (2001). Telomerase activity and patient survival after surgery for gastric and oesophageal cancer. *Eur. J. Gastroenterol. Hepatol., 13(8)*, 903-908.

Okayasu, I., Mitomi, H., Yamashita, K., Mikami, T., Fujiwara, M., Kato, M. & Oshimura, M. (1998). Telomerase activity significantly correlates with cell differentiation, proliferation and lymph node metastasis in colorectal carcinomas. *J. Cancer Res. Clin. Oncol., 124(8)*, 444-449.

Yoshida, R., Kiyozuka, Y., Ichiyoshi, H., Senzaki, H., Takada, H., Hioki, K. & Tsubura, A. (1999). Change in telomerase activity during human colorectal carcinogenesis. *Anticancer Res.*, *19(3B)*, 2167-2172.
Boldrini, L., Faviana, P., Gisfredi, S., Zucconi, Y., Di Quirico, D., Donati, V., Berti, P., Spisni, R., Galleri, D., Materazzi, G., Basolo, F., Miccoli, P., Pingitore, R. & Fontanini, G. (2002). Evaluation of telomerase mrna (htert) in colon cancer. *Int. J. Oncol.*, *21(3)*, 493-497.
Naito, Y., Takagi, T., Handa, O., Ishikawa, T., Matsumoto, N., Yoshida, N., Kato, H., Ando, T., Takemura, T., Itani, K., Hisatomi, H., Tsuchihashi, Y. & Yoshikawa, T. (2001). Telomerase activity and expression of telomerase rna component and catalytic subunits in precancerous and cancerous colorectal lesions. *Tumour. Biol.*, *22(6)*, 374-382.
Niiyama, H., Mizumoto, K., Sato, N., Nagai, E., Mibu, R., Fukui, T., Kinoshita, M. & Tanaka, M. (2001). Quantitative analysis of htert mrna expression in colorectal cancer. *Am. J. Gastroenterol.*, *96(6)*, 1895-1900.
Shoji, Y., Yoshinaga, K., Inoue, A., Iwasaki, A. & Sugihara, K. (2000). Quantification of telomerase activity in sporadic colorectal carcinoma: Association with tumor growth and venous invasion. *Cancer*, *88(6)*, 1304-1309.
Hisatomi, H., Nagao, K., Kanamaru, T., Endo, H., Tomimatsu, M. & Hikiji, K. (1999). Levels of telomerase catalytic subunit mrna as a predictor of potential malignancy. *Int. J. Oncol.*, *14(4)*, 727-732.
Kishimoto, K., Fujimoto, J., Takeuchi, M., Yamamoto, H., Ueki, T. & Okamoto, E. (1998). Telomerase activity in hepatocellular carcinoma and adjacent liver tissues. *J Surg. Oncol.*, *69(3)*, 119-124.
Suda, T., Isokawa, O., Aoyagi, Y., Nomoto, M., Tsukada, K., Shimizu, T., Suzuki, Y., Naito, A., Igarashi, H., Yanagi, M., Takahashi, T. & Asakura, H. (1998). Quantitation of telomerase activity in hepatocellular carcinoma: A possible aid for a prediction of recurrent diseases in the remnant liver. *Hepatology*, *27*, 402-406.
Pearson, A. S., Chiao, P., Zhang, L., Zhang, W., Larry, L., Katz, R. L., Evans, D. B. & Abbruzzese, J. L. (2000). The detection of telomerase activity in patients with adenocarcinoma of the pancreas by fine needle aspiration. *Int. J. Oncol.*, *17(2)*, 381-385.
Hara, T., Noma, T., Yamashiro, Y., Naito, K. & Nakazawa, A. (2001). Quantitative analysis of telomerase activity and telomerase reverse transcriptase expression in renal cell carcinoma. *Urol. Res.*, *29(1)*, 1-6.
Paradis, V., Bieche, I., Dargere, D., Bonvoust, F., Ferlicot, S., Olivi, M., Lagha, N. B., Blanchet, P., Benoit, G., Vidaud, M. & Bedossa, P.

(2001). Htert expression in sporadic renal cell carcinomas. *J. Pathol.*, *195(2)*, 209-217.

De Kok, J. B., Schalken, J. A., Aalders, T. W., Ruers, T. J., Willems, H. L. & Swinkels, D. W. (2000). Quantitative measurement of telomerase reverse transcriptase (htert) mrna in urothelial cell carcinomas. *Int. J. Cancer*, *87(2)*, 217-220.

Nakanishi, K., Kawai, T., Hiroi, S., Kumaki, F., Torikata, C., Aurues, T. & Ikeda, T. (1999). Expression of telomerase mrna component (htr) in transitional cell carcinoma of the upper urinary tract. *Cancer*, *86(10)*, 2109-2116.

Lancelin, F., Anidjar, M., Villette, J. M., Soliman, A., Teillac, P., Le Duc, A., Fiet, J. & Cussenot, O. (2000). Telomerase activity as a potential marker in preneoplastic bladder lesions. *BJU Int.*, *85(4)*, 526-531.

Engelhardt, M., Albanell, J., Drullinsky, P., Han, W., Guillem, J., Scher, H. I., Reuter, V. & Moore, M. A. (1997). Relative contribution of normal and neoplastic cells determines telomerase activity and telomere length in primary cancers of the prostate, colon, and sarcoma. *Clin. Cancer Res.*, *3(10)*, 1849-1857.

Bonatz, G., Frahm, S. O., Klapper, W., Helfenstein, A., Heidorn, K., Jonat, W., Krupp, G., Parwaresch, R. & Rudolph, P. (2001). High telomerase activity is associated with cell cycle deregulation and rapid progression in endometrioid adenocarcinoma of the uterus. *Hum. Pathol.*, *32(6)*, 605-614.

Sangiorgi, L., Gobbi, G. A., Lucarelli, E., Sartorio, S. M., Mordenti, M., Ghedini, I., Maini, V., Scrimieri, F., Reggiani, M., Bertoja, A. Z., Benassi, M. S. & Picci, P. (2001). Presence of telomerase activity in different musculoskeletal tumor histotypes and correlation with aggressiveness. *Int J Cancer*, *95(3)*, 156-161.

Schneider-Stock, R., Jaeger, V., Rys, J., Epplen, J. T. & Roessner, A. (2000). High telomerase activity and high htrt mrna expression differentiate pure myxoid and myxoid/round-cell liposarcomas. *Int J Cancer*, *89(1)*, 63-68.

Wurl, P., Kappler, M., Meye, A., Bartel, F., Kohler, T., Lautenschlager, C., Bache, M., Schmidt, H. & Taubert, H. (2002). Co-expression of survivin and tert and risk of tumour-related death in patients with soft-tissue sarcoma. *Lancet*, *359(9310)*, 943-945.

Brinkschmidt, C., Poremba, C., Christiansen, H., Simon, R., Schafer, K. L., Terpe, H. J., Lampert, F., Boecker, W. & Dockhorn, D. B. (1998). Comparative genomic hybridization and telomerase activity analysis identify two biologically different groups of 4s neuroblastomas. *Br J Cancer*, *77(12)*, 2223-2229.

Dome, J. S., Chung, S., Bergemann, T., Umbricht, C. B., Saji, M., Carey, L. A., Grundy, P. E., Perlman, E. J., Breslow, N. E. & Sukumar, S. (1999). High telomerase reverse transcriptase (htert) messenger rna level correlates with tumor recurrence in patients with favorable histology wilms' tumor. *Cancer Res.*, *59(17)*, 4301-4307.

Dome, J. S., Bockhold, C. A., Li, S. M., Baker, S. D., Green, D. M., Perlman, E. J., Hill, D. A. & Breslow, N. E. (2005). High telomerase rna expression level is an adverse prognostic factor for favorable-histology wilms' tumor. *J Clin Oncol*, *23(36)*, 9138-9145.

Index

A

abnormalities, 108, 118
accuracy, 5, 114, 119, 135
acetylation, ix, 59, 60, 61, 67, 68, 70, 77
acid, 12, 41, 53, 62, 72, 73, 74, 75
acromegaly, 84, 86
ACTH, 86
activators, 70
active site, 47
activity level, 66, 85, 112, 131
acute, 90
adaptation, 18
adenocarcinoma, 112, 121, 122, 137, 143
adenoma, 84, 85, 86, 92, 94, 115, 121, 131, 134, 136
adriamycin, 48
adult, 3, 12, 19, 32, 33, 34, 43, 103, 107, 111, 115, 118
adult stem cells, 19
adult tissues, 3
age, viii, 2, 12, 21, 22, 36, 84
agent, 48, 68, 90
aggressive behavior, 83, 85
aggressiveness, x, 82, 85, 115, 120, 133, 144
aging process, 96, 100
aid, 143
AKT, 22
alpha, 4, 28, 72, 75
ALT, 56, 83, 107, 108, 110, 115, 123
alternative, 18, 23, 38, 83, 93, 98, 107, 110, 115, 120, 123
amenorrhea, 84
amyloid, 9, 12, 30
amyloid beta, 9, 30
anaemia, 45
angiogenesis, x, 6, 41, 63, 76, 82, 86, 87, 115, 131
angiogenic, 63
animals, x, 12, 21, 22, 40, 96
antagonism, 65
antagonist, 89
anti-angiogenic, 75
anti-angiogenic agents, 75
anti-apoptotic, 9, 10, 14, 25, 63
antibody, 86
anticancer, vii, ix, 48, 59, 60, 62, 63, 67, 70, 71, 72, 73, 128
anticancer drug, 48, 62, 73
antigen, 13, 86, 107, 111, 112, 133
antioxidant, 6, 18
antisense, 30, 41, 46
anxiety, 12, 13, 24
APC, 73
apoptosis, viii, 2, 3, 6, 7, 8, 9, 10, 11, 12, 13, 14, 15, 18, 22, 25, 26, 27, 30, 31, 34, 41, 47, 49, 60, 61, 63, 67, 68, 70, 72, 73, 74, 75, 78, 79, 83, 98
apoptosis resistance, 10

Index

apoptotic, 8, 9, 13, 14, 16, 41, 60, 63, 68
application, xi, 88, 106, 112, 120, 136
arrest, 28, 34, 38, 40, 41, 60, 62, 63, 67, 70, 72, 73, 83, 90, 107
ascites, 124, 140
Asian, 89
aspiration, 106, 121, 125, 136, 143
assessment, 114
asthma, 133
astrocytoma, 115
ataxia, 43
athletes, 22
ATM, 47, 48, 49, 90
ATP, 7
autoimmunity, 120
autosomal recessive, 45

B

bacterial, 15
BAL, 116, 119, 125
barrier, 21, 90
Bcl-2, 10, 26, 63
behavior, x, 33, 53, 82, 85, 87, 131
beneficial effect, 22
benefits, 121
benign, x, 82, 84, 85, 92, 109, 110, 115, 118, 120, 121, 122, 125, 129, 133, 134, 135
benign prostatic hyperplasia, 135
benign tumors, x, 82, 85, 110, 125, 134
bile, 109, 120, 134, 135
bile duct, 109, 120
biliary tract, 135
binding, 7, 15, 22, 26, 34, 41, 42, 44, 47, 55, 64, 65, 66, 67, 75, 77, 84, 93, 97, 98, 99, 100, 106
biogenesis, 24, 32, 42, 53
biological behavior, x, 82, 85, 87
biological processes, ix, 36, 43, 81
biomarker, xi, 83, 93, 106, 112, 114, 120, 121, 123, 138
biomedical applications, 33
biopsy, 89, 119, 120, 121, 122, 123, 124, 133, 135, 136

bladder, 62, 73, 83, 108, 120, 135, 143
bladder cancer, 62, 73, 120, 135
BLM, 101
blocks, 45
blood, 124, 125
blot, 19
borderline, 108, 123
bovine, 29
bowel, 108, 128
brain, viii, x, 2, 9, 12, 13, 27, 31, 68, 69, 82, 83, 85, 89, 92, 107, 109, 115, 118, 131
brain development, 31
brain functions, 12
brain injury, 27
brain tumor, x, 82, 83, 85, 89, 92, 107, 115, 118, 131
breast cancer, 29, 39, 40, 60, 118, 121, 124, 125, 132, 136, 137, 140, 141
breast mass, 136
broad spectrum, 62
bronchial epithelium, 141
bronchoalveolar lavage, 116, 119, 125
buffer, 69, 112
bulbs, 12
bypass, 108

C

calcium, 15, 16
caloric restriction, x, 96, 101, 102
cAMP, 91
cancer progression, xi, 106, 107, 109, 114
cancer screening, 120
cancer stem cells, 20
cancer treatment, 72
cancerous cells, 70, 125
carcinogen, 39
carcinogenesis, xi, 32, 106, 107, 108, 112, 113, 114, 119, 123, 126, 128, 130, 139, 142
carcinoma, 73, 75, 86, 88, 91, 108, 112, 114, 117, 119, 120, 121, 123, 124, 127, 128, 129, 133, 134, 135, 136, 137, 138, 140, 142, 143
cardiomyocytes, 22

Index

cardiovascular system, 22
cartilage, 45, 56
caspase, 9, 10, 68, 75
cassettes, 50
catalase, 18, 19, 31
catalytic activity, viii, 2, 4, 10, 11, 12, 20, 21, 24
causality, 22
CDK, 78
CDK4, 72
CDK9, 61
cell cycle, x, 6, 29, 40, 43, 46, 63, 64, 70, 73, 82, 85, 87, 98, 143
cell cycle molecules, x, 82, 87
cell death, 9, 11, 12, 13, 26, 30, 32, 63, 74, 75
cell differentiation, 6, 7, 141, 142
cell division, 38, 60, 82, 106, 108
cell fate, 100
cell growth, 31, 41, 44, 45, 55, 57, 75, 90, 101
cell line, 6, 9, 33, 38, 41, 42, 46, 48, 53, 69, 84, 89, 107, 132
cell surface, 44
central nervous system, 89, 140
cerebellum, 12
cervical cancer, 68, 79, 123, 130
cervical dysplasia, 123, 138
cervix, 114, 123, 129, 138
chaperones, 61
chemoprevention, 114
chemotherapeutic agent, 62, 75
chemotherapeutic drugs, 9
chemotherapy, 72, 117, 118, 130, 133
childhood, 107, 109, 115, 118
chromatin, vii, viii, 2, 3, 7, 8, 20, 26, 28, 29, 44, 51, 56, 60, 62, 67, 69, 70, 71, 72, 79, 139
chromosomal abnormalities, 108
chromosomal instability, 19, 33, 68
chromosome, ix, 4, 5, 7, 28, 36, 38, 43, 54, 67, 81, 87, 88, 101, 106, 126
cigarette smoke, 141
cigarette smokers, 141
cilia, 62, 68

cis, 65, 66, 97, 100
cisplatin, 30
classes, 5, 61, 62
classical, 24
classification, 91, 94
cleavage, 30, 75
click chemistry, 82
clinical symptoms, 87
clinical trial, 62, 71
clinics, 125
cloning, 101
c-Myc, 61, 64, 65, 66, 67, 68, 70, 71, 76, 77, 78, 98, 100, 102
CNS, 69
coding, 6, 19, 24, 39, 93
codon, 65
cofactors, 72, 77
collaboration, 71
colon, 44, 73, 84, 91, 108, 110, 120, 124, 142, 143
colon cancer, 73, 84, 91, 142
colorectal cancer, 83, 88, 115, 117, 118, 130, 131, 142
combination therapy, 62
communication, x, 82
community, 15
complementary DNA, 4
complexity, 43
complications, 121
components, vii, 1, 3, 6, 11, 12, 13, 15, 23, 25, 32, 41, 45, 50, 92, 111, 123, 133, 138
composition, iv, 4, 45, 52
compounds, 62, 70, 97
comprehension, 44
concentration, 48, 98
configuration, 7, 8, 67
conflict, 67
consensus, 98, 99
construction, 96
consumption, 15
contamination, 110, 119, 124
context-dependent, 61
control, 13, 29, 52, 64, 69, 96, 97, 100, 106, 110
conversion, 40, 53

correlation, vii, 1, 8, 86, 123, 130, 141, 144
cortex, 18
covalent, 60
cranial nerve, 84
CT-scanning, 45
cutaneous T-cell lymphoma, 62, 74
cycles, 17
cyclin-dependent kinase inhibitor, 98
cyclins, 6
cytochrome, 9
cytokines, 85, 91
cytologic examination, 139, 140
cytology, 106, 121, 124
cytoplasm, 10, 13, 98
cytotoxic, 62

D

data analysis, 16
de novo, 4, 28, 64, 101
death, 3, 10, 13, 40, 62, 63, 73, 75, 83, 92, 144
decisions, 118
defects, 11, 20, 56, 62, 84
defense, 33
defense mechanisms, 33
deficiency, 12, 18, 31, 58
degenerate, 65
degradation, 48, 82, 96, 107, 111
delivery, 55
demographic characteristics, 92
density, 86
dental plaque, 119
deprivation, x, 96
deregulation, 143
destruction, 48
detection, x, xi, 50, 82, 93, 105, 110, 111, 112, 113, 114, 115, 119, 120, 121, 122, 123, 124, 125, 127, 129, 135, 139, 140, 143
developing brain, 30
diagnostic markers, 122
dicentric chromosome, 4
differential diagnosis, 120, 121, 123, 133

differentiation, 6, 7, 29, 33, 39, 52, 60, 65, 70, 74, 77, 90, 141, 142
direct action, 41
discrimination, 121
diseases, 4, 107, 108, 120, 122, 143
disorder, 45, 47
displacement, 36, 67, 71
dissociation, 50, 68
distribution, 18, 123
division, vii, ix, 38, 60, 81, 82, 106
DNA damage, viii, 2, 3, 5, 7, 21, 27, 28, 29, 30, 32, 37, 41, 43, 47, 48, 49, 50, 51, 54, 57, 67, 68, 97, 98
DNA polymerase, viii, 2, 23, 37, 49, 53, 55, 96, 106
DNA repair, x, 5, 7, 29, 37, 48, 49, 50, 82, 95, 96
double helix, ix, 81, 82
dream, 3
drug resistance, 48
drug treatment, 9, 10
drugs, 9, 10, 15, 97, 100, 118
duplication, 45
dysplasia, 108, 113, 114, 123, 129, 130, 138

E

effusion, 116, 124
ELISA, 17
elongation, vii, viii, 2, 5, 28, 40, 54, 84, 127
embryo, 4, 54
embryonic development, 20, 21
embryonic stem, 8, 19, 33
embryonic stem cells, 8, 19, 33
encoding, 60, 64, 96, 98
endocrine, 84, 128
endometrial cancer, 74
endonuclease, 49
endosperm, 4
endothelial cell, 14, 22, 27, 63, 86, 108, 127, 131
endothelial progenitor cells, 34
endurance, 22
energy, 87
environment, 98

environmental stimuli, 61
enzymatic, 31, 47, 85
enzymatic activity, 31, 47
enzymes, 5, 18, 47, 61
ependymal, 62, 68
ependymoma, 69, 89, 92, 131
epidermal growth factor, 39
epidermal growth factor receptor, 39
epidermal stem cells, 19, 43, 44
epidermis, 141
epigenetic, ix, 59, 61, 70, 72, 73
epigenetic mechanism, 70
epigenetics, 71
epithelia, 113, 122, 130
epithelial cell, 6, 29, 30, 38, 53, 57, 109, 111, 124
epithelial stem cell, 57
epithelium, 110, 130, 141
erosion, 48, 58, 67, 106
erythrocytes, 119
Escherichia coli, 90
esophageal cancer, 119, 120, 122
esophagus, 114, 122, 129, 130
estrogen, 132
eukaryotes, 13, 36
eukaryotic cell, 60, 82
evolution, vii, 1, 5, 50, 56, 61, 73, 127
excision, 49, 55
excitotoxins, 12
exclusion, 8, 10, 15
excretion, 120
exercise, 22, 34
exposure, 40, 48, 49, 63, 68, 69

F

failure, 47, 71
false negative, 119
false positive, 111, 120, 124
family, x, 27, 39, 41, 43, 44, 61, 73, 96
feedback, 24, 43, 46, 55
fibroadenoma, 121
fibroblast, 103
fine needle aspiration, 106, 121, 125, 143
first generation, 11, 18, 21

fission, 25, 126
fitness, 21
flexibility, 5
fluid, 124, 136
fluorescence, 87, 128
fluorescence in situ hybridization, 128
fluorophores, 87
FNA, 116, 121, 125
follicle, 43, 141
follicular, 121, 136
Food and Drug Administration (FDA), 62
fractionation, 17
free radical, 103
fusion, 96

G

galactorrhea, 84
gall bladder, 135
gastric, 48, 56, 115, 122, 130, 137, 142
gastrointestinal, 124
gastrointestinal tract, 124
gender, 84
gene amplification, 128
gene promoter, 66, 77, 102
generation, 5, 11, 14, 18, 19, 21, 63, 98
genetic disease, 4
genetic instability, 106
genetics, iv, 72
genome, x, 5, 8, 15, 36, 44, 47, 48, 56, 60, 68, 70, 95
genomic, 3, 5, 29, 30, 47, 48, 50, 51, 63, 78, 99, 101, 106, 108, 144
genomic instability, 101, 108
genotoxic, 10, 31, 49
germ cells, 107, 109
gigantism, 84
gland, 87, 121
glial, 69, 83, 89
glioblastoma, 30, 60, 62, 83, 84, 89, 90, 115
glioblastoma multiforme, 60, 84, 90, 115
glioma, 74, 89, 90, 93, 115
glucose, x, 21, 96, 97, 98, 100, 102
glucose tolerance, 21
glycolysis, 7

gold, 93
gold standard, 93
grades, 115
groups, 3, 7, 9, 15, 18, 65, 144
growth factor, 34, 39, 44, 53, 57, 58, 66, 75, 77, 78
growth hormone, 22, 85, 86, 94
growth inhibition, 26, 39, 55, 57
growth rate, 6, 38, 39
gut, 122

H

hair follicle, 20, 33, 43, 56, 141
head and neck cancer, 141
healing, 4, 5, 7, 28, 40
health, 22
heart, 15, 18, 21, 34
heart failure, 34
heat, 69, 87, 112
helix, ix, 64, 81, 82
hematologic, 52
hematopoietic, 128
hematopoietic progenitor cells, 128
hepatocellular, 123, 137, 140, 142, 143
hepatocellular carcinoma, 123, 137, 140, 142, 143
hepatocytes, 122
heterochromatic, 96
heterochromatin, 101
heterodimer, 37, 42
heterogeneous, 41, 45, 52, 53, 57, 84, 139
heterozygosity, 133
high risk, 109, 113, 123, 125
high-level, 123
hippocampus, 12
histological, 86, 123
histological markers, 130
histology, 115, 118, 138, 144
histone, vii, ix, 7, 59, 60, 61, 62, 63, 65, 68, 69, 70, 71, 72, 73, 74, 75, 76, 77, 78, 79, 90
Histone deacetylase, 72, 73, 74, 75, 76, 78, 79

histone deacetylase inhibitors, ix, 59, 72, 73, 75
holoenzyme, 64
homeostasis, 67, 141
homology, 61
hormone, 6, 22, 61, 84, 85, 86, 94
HPV, 123
H-ras, 107
human brain, 72, 89, 90, 131
human embryonic stem cells, 33
human ES, 29
human ESC, 29
human leukemia cells, 75
human papillomavirus, 130, 138
humanity, 3
hybrid, 82
hybridization, 106, 111, 125, 128, 134, 136, 144
hydrogen, 26
hydrogen peroxide, 26
hyperplasia, 135
hypersensitivity, 49
hypertension, 84
hypoplasia, 45, 56
hypothesis, 20, 40, 43, 48, 60, 107, 127
hypoxia, 63, 71, 75
hypoxia-inducible factor, 75
hysterectomy, 123, 138

I

ice, 11, 12, 21
identification, 62, 77
IGF, 22, 102
IGF-1, 22
IGF-I, 102
image analysis, 138
imaging, 16, 91
immortal, 31, 39, 40, 52, 53, 64, 76, 88, 107, 126
immortality, vii, xi, 2, 3, 53, 60, 76, 89, 105
immunity, 45
immunocompromised, 40
immunohistochemical, 125, 138
immunohistochemistry, 85, 106, 111, 125

immunoprecipitation, 42
in situ, xi, 12, 32, 106, 108, 109, 110, 111, 114, 119, 121, 125, 128, 129, 135, 136
in situ hybridization, 106, 111, 125, 128, 136
in transition, 143
in vitro, 4, 6, 9, 11, 18, 23, 29, 31, 42, 46, 49, 52, 60, 63, 66, 74, 76, 90, 107, 112, 131
in vivo, 4, 6, 7, 11, 12, 18, 23, 40, 42, 44, 46, 49, 56, 60, 63, 74, 90, 92, 112
inactive, 8, 10, 15, 20, 39, 44, 46
incidence, 108
indication, 83, 121
indicators, xi, 106, 132
indices, 85
induction, viii, x, 2, 6, 8, 9, 10, 15, 19, 20, 23, 33, 38, 40, 41, 44, 48, 49, 50, 56, 63, 68, 75, 78, 93, 96, 97
infinite, vii, 2
inflammatory, 21, 108, 121, 128
inflammatory bowel disease, 108, 128
inflammatory cells, 121
inhibition, vii, 10, 13, 26, 30, 31, 39, 41, 45, 55, 57, 60, 62, 66, 68, 69, 70, 78, 83, 84, 88, 90, 91
inhibitor, 9, 10, 26, 48, 72, 73, 74, 75, 77, 78, 79, 90, 98
inhibitory, 41, 67, 73
inhibitory effect, 67, 73
initiation, ix, 59, 61, 62, 65, 66, 71, 77
injury, iv, 12, 122
insight, viii, 2
instability, x, 88, 96, 106, 108, 126, 127
insulin, 34, 98
insulin signaling, 98
insulin-like growth factor, 34
insults, 49, 113
integration, 89
integrity, 5
interaction, vii, viii, 2, 5, 7, 9, 10, 42, 43, 78
interference, 7, 8, 24, 25, 26, 46, 49, 54, 55
internalization, 53
interphase, 84
intestine, 122, 128

intracranial, 84, 86, 89, 118, 131
intracranial tumors, 86, 118
intrinsic, 63, 64, 75
invasive, 85, 91, 121, 136
Investigations, 123
iodine, 129
ionizing radiation, 75, 101
ions, 23
irradiation, 43, 48, 49
ischemia, 27
ischemic, 9, 12
ischemic brain injury, 9
isoforms, 94, 101

K

karyotypes, 108
keratinocytes, 34, 40, 53, 57
kinase, 14, 27, 34, 41, 42, 43, 47, 51, 57, 58
kinase activity, 42, 51
kinetics, 49
knockout, 11, 15, 18, 20, 21, 22, 44

L

labeling, 129
laser, 48
lens, 29
lentiviral, 55
lesions, 108, 113, 119, 121, 122, 123, 129, 130, 133, 137, 138, 142, 143
leucine, 64
leukemia, 75, 84
leukemia cells, 75
leukemic, 78
leukocytes, 22, 34, 52, 119
life span, x, 21, 31, 96, 97, 98, 127
lifetime, 40
ligand, 10, 75
linear, 36, 57, 107
links, 32, 53
lipopolysaccharide, 22
lipoxygenase, 73
liver, 55, 68, 78, 118, 122, 125, 142, 143

liver cancer, 78
liver cells, 68
liver disease, 125
liver metastases, 118
localization, 26, 32, 50, 51
location, 70
locus, 50
longevity, 102
losses, 108
low-level, 121, 123
luciferase, 39
luminal, 120, 134
lung, 15, 39, 62, 65, 73, 75, 113, 114, 117, 119, 124, 129, 130, 131, 133, 134, 139, 141
lung cancer, 73, 113, 114, 117, 119, 130, 131, 133, 134, 139, 141
lung disease, 134
lymph, 139, 142
lymph node, 139, 142
lymphocytes, xi, 13, 22, 105, 107, 109, 110, 111, 120, 121, 124, 128, 133
lymphoid, 66
lymphoid cells, 66
lymphoma, 62, 74, 117, 140
lysine, 61, 69, 70, 71, 79

M

machinery, 13, 96
magnetic, iv, 91
magnetic resonance, 91
magnetic resonance imaging, 91
maintenance, vii, viii, ix, x, 2, 18, 24, 31, 35, 38, 41, 42, 44, 47, 60, 67, 70, 81, 84, 89, 90, 92, 93, 96, 101, 107
maize, 5
malignant, x, 30, 82, 85, 89, 92, 108, 109, 110, 111, 113, 114, 115, 117, 118, 119, 120, 121, 123, 124, 129, 133, 137, 138, 139, 140
malignant cells, 110
malignant melanoma, 123, 138
malignant tumors, 109, 111, 115, 118
mammalian, 37, 53, 78, 100

mammalian cell, 7, 8, 64, 82, 90, 98
mammals, 5, 13, 15, 21, 36, 37, 97, 98
management, 91
manganese, 23
manipulation, 66
MAPK, 92
matrix, 16, 96
maturation, 73
MDR, 48
measurement, 122, 124, 143
median, 21
medicine, 3
medulloblastoma, 60, 69, 74, 75, 90
MEF, 11
melanin, 6
melanocytic lesions, 138
melanoma, 6, 7, 26, 60, 62, 123, 138
memory, 133
mesoderm, 21, 58
metabolism, iv, ix, 6, 25, 36, 42, 50
metal ions, 23
metals, 23
metastasis, 6, 26, 41, 117, 118, 142
metastasize, 124
metastatic, x, 7, 81
methylation, 60, 73, 139
MIB-1, 86
mice, 9, 11, 12, 13, 15, 18, 19, 20, 21, 31, 32, 34, 36, 40, 44, 51, 53, 57, 74, 107, 108, 127
microarray, 6
microenvironment, 71
microscopy, 16
microtubule, 74
middle-aged, 22
migration, 32
mimicking, 48
misleading, 24
mitochondria, viii, 2, 3, 8, 9, 10, 13, 15, 17, 18, 24, 25, 27, 33, 45
mitochondrial damage, 75
mitochondrial death, 13
mitochondrial DNA, 16, 27
mitochondrial membrane, 6, 12, 15, 16
mitogen, 91

mitogen-activated protein kinase, 91
mitogenic, 38
mitosis, x, 82, 84
mitotic, 12, 53, 85
MnSOD, 18
modalities, 62
models, 11, 25, 40, 60
modulation, 8, 39, 41, 70, 78, 102
molecular biology, x, 82, 87
molecular mechanisms, viii, 2, 3, 50, 73, 76
molecules, x, 10, 36, 45, 46, 82, 85, 86, 87, 91, 125
monoclonal, 111
monoclonal antibodies, 111
mononuclear cell, 124
monotherapy, 62
morphological, 121
motor neurons, 11
mouse, 6, 8, 12, 15, 22, 39, 40, 42, 45, 54, 58, 103, 107, 127
mouse model, 40
mRNA, xi, 30, 45, 54, 68, 78, 83, 89, 90, 106, 111, 114, 115, 116, 117, 118, 120, 121, 122, 123, 124, 125
mtDNA, 15, 26
mucosa, 113, 122, 129, 133
mucus, 119
multicellular organisms, 3
multidrug resistance, 48, 56
multiple myeloma, 62
multipotent, 20, 43
multipotent stem cells, 20, 43
murine cell, 53
musculoskeletal, 144
mutant, 10, 15, 25, 44, 46, 49, 55, 74, 84
MYC, 20, 76
myelomas, 62
myocyte, 34
MyoD, 61

N

NAD, 61, 69, 97
natural, 13, 72, 97, 133
neck, 117, 119, 120, 122, 133, 137, 141
neck cancer, 117, 141
necrosis, 30, 75, 125
needle aspiration, 106, 121, 125, 136, 143
negative regulatory, 66
neoplasias, 83, 84, 108
neoplasm, 128
neoplastic, 51, 65, 68, 71, 136, 138, 143
neoplastic cells, 65, 68, 71, 143
nerve, 11
nervous system, 117, 140
network, 39, 64, 76
neural stem cell, 12, 33
neuroblastoma, 89, 93, 115, 128, 131, 132
neurogenesis, 13
neuronal cells, 11, 12
neurons, 9, 11, 12, 13, 30
neurotoxicity, 12, 27
next generation, 70
NMDA, 9, 11, 12, 27
N-methyl-D-aspartic acid, 12
NMR, 82
Nobel Prize, vii, 1, 3
nodes, 139
non-invasive, 84
non-small cell lung cancer, 114, 117, 131, 141
NOS, 22
notochord, 45
N-terminal, 15
nuclear, 7, 8, 10, 13, 14, 15, 27, 32, 33, 41, 45, 52, 53, 57, 65, 86, 93, 96, 101
nucleic acid, 41, 53
nucleoplasm, 14
nucleoprotein, 42, 67, 106
nucleotides, 5

O

observations, x, 8, 13, 43, 48, 96, 98, 100, 107, 108
olfactory, 12, 13, 32, 83, 89
olfactory bulb, 12, 13
oligonucleotides, 67
oncogene, 26, 29, 30, 31, 53, 73, 75, 76, 126, 127, 136, 138

oncology, 128
oncoproteins, 67
oral, 32, 73, 74, 114, 119, 130, 133, 140
oral cavity, 140
oral squamous cell carcinoma, 73
organ, 109, 123
organelles, 13
oropharynx, 117, 140
osteosarcoma, 75, 115
ovarian cancer, 30, 124, 140
ovary, 109, 124
oxidative, viii, 2, 6, 9, 10, 14, 15, 17, 18, 19, 23, 25, 27, 31, 49, 63, 98
oxidative damage, 18, 49
oxidative stress, viii, 2, 6, 9, 10, 14, 15, 17, 18, 19, 23, 25, 27, 31, 63, 98
oxygen, 15, 18, 32
oxygen consumption, 15

P

paclitaxel, 74
pancreas, 108, 114, 120, 123, 128, 134, 143
pancreatic, 60, 62, 112, 120, 129, 134
pancreatic cancer, 60, 129, 134
pancreatitis, 134
paraffin-embedded, 111, 112, 138
Parkinson, 10, 31
pathogenesis, 47, 60, 94
pathways, viii, 2, 7, 8, 20, 22, 24, 26, 29, 34, 37, 38, 43, 44, 50, 60, 62, 63, 75, 92
patients, x, xi, 6, 57, 62, 84, 85, 89, 90, 96, 98, 106, 109, 113, 115, 118, 119, 120, 121, 122, 124, 125, 132, 135, 138, 139, 140, 143, 144
PCR, 83, 85, 88, 93, 99, 110, 111, 119, 120
pediatric, 89, 115, 128, 131, 132
peptide, 53
perception, 12
peripheral blood, 124
peripheral blood mononuclear cell, 124
peritoneal, 124
peritoneal lavage, 124
peritoneum, 124
permit, 48, 108

peroxide, 27
perturbations, 47
phenotype, 40, 48, 60, 64, 130, 139
phenotypic, 13, 70
pheochromocytoma, 12
phosphorylation, 5, 7, 13, 27, 32, 34, 42, 43, 49, 50, 54, 57, 58, 60
phylogenetic, 5, 73
physical activity, 22
physical exercise, 22, 34
physiological, 11, 42
pigments, 112
pituitary, x, 82, 84, 85, 86, 87, 91, 92, 94, 115, 131
pituitary gland, 85, 87
pituitary tumors, 85, 91
plants, 5, 13
plaque, 119
plasmid, 57
plasticity, viii, 2, 23
play, vii, ix, x, 21, 36, 37, 38, 41, 43, 47, 50, 63, 81, 95, 96, 98
pleural, 124, 139
pleural effusion, 124, 139
plurality, 62
pluripotency, 19
point mutation, 101, 134
polymerase, viii, 2, 23, 27, 46, 47, 49, 55, 83, 85, 87, 88, 106, 135, 136
polymerase chain reaction, 85, 88, 135, 136
polyp, 88
poor, 83, 114, 115, 131, 132, 139
population, 36, 91
pores, 14
postoperative, 118
post-translational, 13
preclinical, 60
prediction, 143
predictors, 132
primary tumor, 83
probe, 16
production, 7, 32, 75, 85
progenitor cells, xi, 8, 12, 19, 20, 21, 56, 71, 105, 109, 110, 122, 128
progesterone, 132

prognosis, vii, ix, 81, 83, 84, 87, 106, 114, 115, 117, 118, 132, 139, 141
prognostic factors, 118
prognostic marker, 83, 85, 113, 115, 117, 118, 125
program, 29, 38, 44, 52, 67
prolactin, 85, 86
promoter, ix, xi, 39, 44, 59, 64, 65, 66, 67, 68, 70, 71, 72, 76, 77, 79, 96, 97, 98, 99, 100, 102
promoter region, xi, 76, 96, 97, 98, 99, 100
promyelocytic, 65, 84
propagation, 69
property, iv, 5, 8
prostate, 31, 68, 75, 78, 83, 93, 108, 114, 121, 135, 136, 143
prostate cancer, 31, 68, 78, 93, 114, 121, 135
prostate carcinoma, 75
prostate gland, 121
prostatectomy, 136
proteases, 15
protection, 9, 10, 13, 48, 52, 58, 108
protective role, 40
protein kinases, 91
protein structure, ix, 81, 82
protein synthesis, 24
protein-protein interactions, 49, 66
proteome, 94
proteomics, 87
protocol, 69, 83, 85, 86, 106, 125, 129, 130
proto-oncogene, 55, 65, 66
public, 3

R

radiation, 75, 101, 140
radiation therapy, 140
radical hysterectomy, 123, 138
radiosensitization, 74
random, 5
range, 36, 40
ras, 107, 134
rat, 62, 68
reactive oxygen, 32, 63, 75, 98

reactive oxygen species, 63, 75, 98
reading, 50
reagents, 82
real time, 120
receptors, 44, 61, 133
reciprocal translocation, 127
recognition, 42, 98
recombination, 50, 55, 84, 88, 90, 91
recovery, 28
recruiting, 66
recurrence, x, 82, 84, 88, 89, 92, 131, 144
redox, 4, 18, 27
regeneration, 122
regional, 123
regression, 75, 115
regrowth, 115
regulation, viii, ix, x, 2, 6, 8, 24, 29, 31, 33, 36, 38, 44, 45, 46, 48, 53, 54, 59, 60, 64, 65, 66, 71, 75, 76, 81, 93, 96, 98, 102, 127
regulators, 61, 68
relapse, 117
relationship, 18, 46, 84, 115
relevance, 6
reliability, 93
remodelling, 8, 42, 51, 60, 70
renal, 18, 62, 143
renal cell carcinoma, 143
repair, viii, x, 2, 5, 7, 29, 37, 42, 47, 48, 49, 50, 54, 55, 56, 82, 84, 88, 90, 91, 95, 96, 100, 141
replication, 37, 52, 58, 82, 84, 96, 106
repression, 24, 60, 61, 63, 64, 65, 66, 67, 70, 71, 76, 77, 79
repressor, 65, 69, 70
Research and Development, 105
resection, 85, 118, 132
reserves, 40
residues, 61
resistance, viii, 2, 9, 10, 11, 12, 22, 31, 40, 48, 56, 57, 62, 98
resolution, 45
respiration, 15
restructuring, 5
Resveratrol, 96, 100

ribosomal, 24, 45
ribosomal RNA, 24
ribosomes, 24
ribozyme, 6
RISC, 46
risk, xi, 106, 109, 113, 114, 117, 118, 122, 123, 125, 130, 144
RNA processing, 24, 45
rodents, 3
ROS, 10, 14, 15, 18, 63, 98
RRM, 42

S

safeguards, 52, 78
SAHA, 62, 75
Salmonella, 90
sample, 112
sarcomas, 107, 115, 132, 138, 139
scientific community, 15
scores, 99
secretion, 85, 120
self-renewal, xi, 19, 105
senescence, ix, x, 3, 7, 14, 22, 23, 27, 30, 32, 34, 38, 39, 47, 51, 67, 69, 72, 78, 81, 83, 93, 95, 96, 98, 100, 106, 107, 127, 128
sensitivity, viii, 2, 3, 8, 9, 11, 12, 15, 19, 48, 62, 110, 114, 119, 120, 121, 123, 124
separation, 124
series, ii, 86
serum, 22, 86, 125, 140
shelter, ix, 36, 37, 56, 67, 88
short-term, 20
shuttles, 3
side effects, 15, 50
signal transduction, 67, 98
signaling, 29, 53, 57, 58, 75, 91, 94, 98, 102
signaling pathway, 29
signalling, viii, 2, 5, 6, 8, 20, 24, 26, 28, 33, 39, 44, 45, 50, 55, 56, 66, 98
silencers, 70
similarity, 47
siRNA, 6, 10, 18, 46
sister chromatid exchange, 84, 93
sites, 4, 5, 7, 47, 48, 49, 50, 61, 64, 65, 66, 99, 100, 124
skin, 8, 20, 34, 40, 43, 53, 57, 123, 138
skin cancer, 123, 138
small intestine, 44
smokers, 113, 117, 119, 133, 141
sodium, 62, 68
sodium butyrate, 62, 68
soft tissue sarcomas, 115, 132, 138, 139
soft tissue tumors, 123, 133, 139
solid tumors, 115
somatic cell, viii, xi, 35, 38, 41, 43, 64, 65, 67, 69, 71, 76, 82, 83, 105, 106, 110
Southeast Asia, 89
soybean, 5
spatial, 137
specialized cells, 109
species, 3, 5, 36, 63, 75, 98
specificity, 120, 121, 124
spectrum, 62
S-phase, 7, 56, 64
sporadic, 88, 136, 142, 143
sputum, 119, 133
squamous cell, 73, 120, 123, 129, 133, 137
squamous cell carcinoma, 73, 120, 123, 129, 133, 137
stability, x, xi, 3, 29, 47, 78, 95, 106
stages, xi, 49, 105, 112, 113, 117, 124
starvation, 102
staurosporine, 11, 14
steady state, 49
Stem cell, 141
stomach, 122
storage, 15, 16
strain, 40, 44
strategies, viii, 2, 62, 63, 67, 71
stress, viii, 2, 6, 8, 9, 10, 11, 14, 15, 17, 18, 19, 22, 25, 27, 29, 30, 31, 33, 39, 58, 63, 98
stromal, 73
stromal cells, 73
students, 71
substances, 119
substrates, 42, 49, 61, 63
subventricular zone, 12, 32

supply, 12
suppression, 6, 7, 12, 13, 66, 67, 68, 93
suppressor, 60, 63, 66, 67, 76
surgery, 86, 142
surgical, 85, 136
surgical resection, 85
surprise, 15
surrogates, 111
surveillance, 87
survival, 6, 9, 11, 18, 21, 30, 31, 34, 56, 61, 62, 88, 89, 90, 98, 115, 131, 132, 142
susceptibility, 3, 30, 40, 75
switching, 23, 43
symptoms, 84, 87
syndrome, x, 47, 84, 96, 101, 102
synergistic, 22, 62, 74
synergistic effect, 22, 74
synthesis, vii, ix, x, 1, 3, 6, 24, 43, 46, 47, 49, 55, 64, 81, 83, 96, 101
systems, 10, 87

T

T cells, 93
T lymphocyte, 32
tandem repeats, 37
target identification, 62
targets, 39, 50, 60, 61, 62, 63, 69, 70, 72, 89, 98
technology, 86, 87, 111
telomere shortening, xi, 23, 27, 30, 31, 40, 41, 49, 56, 68, 106, 108, 127
temporal, 63, 108
testis, 109
tetracycline, 44
TGF, 39, 66, 78
thawing, 17
therapeutic agents, 72
therapeutics, 54, 71, 78
therapy, 30, 50, 62, 69, 70, 74, 140
thioredoxin, 75
threshold, 38
threshold level, 38
thyroid, 109, 110, 118, 121, 129, 133, 136, 141

thyroid carcinoma, 129
thyroid gland, 121
thyroiditis, 121
time, 7, 11, 20, 48, 53, 86, 87, 88, 93, 120, 135, 136
time bomb, 53
tissue, 4, 15, 18, 20, 22, 36, 40, 62, 85, 86, 112, 115, 118, 119, 121, 122, 123, 132, 133, 134, 136, 138, 139, 144
tobacco, 113
tolerance, 21
toxicity, 62, 68
traffic, 13
training, 23
transcription factor, 44, 45, 50, 52, 61, 64, 65, 70, 76, 78, 87, 97, 98, 99, 100
transcriptional, viii, xi, 2, 8, 20, 21, 22, 24, 29, 39, 44, 45, 52, 60, 64, 65, 66, 68, 70, 71, 76, 77, 96, 97, 102
transduction, 39
transfection, 100
transformation, xi, 30, 32, 38, 40, 41, 45, 51, 62, 65, 76, 92, 106, 108, 109, 112, 113
transforming growth factor, 39, 53, 57, 77
transgenic, 11, 12, 20, 21, 34, 40, 44, 53
transgenic mice, 34, 40, 53
transgenic mouse, 44
transition, 20, 43, 108
transitional cell carcinoma, 135, 143
translation, 17, 24
translocation, 10, 32, 97, 127
transport, 17
triage, 123, 138
trichostatin, 62, 73, 77, 78
trichostatin A, 62, 73, 77, 78
triggers, 27, 47, 78
TSA, 62, 65, 67, 68, 69, 73
tuberculosis, 124
tumor cells, 49, 56, 74, 76, 78, 86, 125
tumor growth, 89, 142
tumor necrosis factor, 75
tumor progression, vii, ix, xi, 81, 85, 106, 113, 114, 119

tumorigenesis, ix, 26, 39, 40, 52, 57, 73, 81, 91, 92, 107, 127
tumour growth, 41
tumour suppressor genes, 63
turnover, 12
tyrosine, 27

U

underlying mechanisms, viii, 2
urinary, 143
urinary tract, 143
urine, 116, 120, 121, 135
uterus, 123, 143
UV irradiation, 43

V

valproic acid, 62, 73, 74
values, 17, 125
variability, 5
variables, x, 82, 85, 86, 87
variation, 89
vascular endothelial growth factor, 58, 63, 75
vector, 39
VEGF expression, 39

vertebrae, 21, 45
vessels, 86
visualization, 48

W

wild type, 11, 12, 22, 49
withdrawal, 12
Wnt signaling, 55
women, 123, 138
workers, 6, 9, 14

X

xenografts, 90

Y

yeast, 4, 5, 13, 23, 25, 28, 32, 33, 56, 57, 61, 97, 126

Z

zinc, 61